Unnaturally
Delicious

Unnaturally
Delicious

HOW SCIENCE AND TECHNOLOGY
ARE SERVING UP SUPER FOODS
TO SAVE THE WORLD

JAYSON LUSK

St. Martin's Press
New York

For Jackson and Harrison

www.stmartins.com

Design by Letra Libre, Inc.

Library of Congress Cataloging-in-Publication Data

Names: Lusk, Jayson, author.

Title: Unnaturally delicious : how science and technology are serving up super foods to save the world / by Jayson Lusk.

Description: First edition. | New York : St. Martin's Press, [2016]

Identifiers: LCCN 2015043832| ISBN 9781250074300 (hardcover) | ISBN 9781466885950 (e-book)

Subjects: LCSH: Agricultural innovations. | Food industry and trade— Technological innovations. | Food supply. | Genetically modified foods.

Classification: LCC S494.5.I5 L87 2016 | DDC 338.1—dc23

LC record available at http://lccn.loc.gov/2015043832

First Edition: March 2016

10 9 8 7 6 5 4 3 2 1

Contents

Acknowledgments

I am indebted to my wife, Christy, for her great ideas, keen editing, and constant encouragement. My friends and colleagues Bailey Norwood, Trey Malone, and Brandon McFadden helped me think through the issues as this book took shape and provided valuable suggestions and advice throughout the process.

I have never before written a book that required so much from others. Fortunately, many people were willing to give their time, insights, and data to help me tell the stories of food and agricultural innovation. I want to thank Abdul Naico and the team at HarvestPlus (including Ekin Birol, Yassir Islam, Erick Boy, Wolfgang Pfeiffer, Amy Saltzman, and Adewale Oparinde) for providing feedback on the topic of biofortification and Ingo Potrykus for taking the time to communicate with me about golden rice. Justin Siegel and Richard Kong gave me feedback on the chapter on synthetic biology and helped me track down their students who were involved in the iGEM competition. I'm grateful to Aaron Cohen, Sara Ritz, Marc So, and Paul Tse, all of who took the time to answer my questions about their iGEM projects. Ian Duncan, Christy Goldhawk, and Tom Silva all provided useful suggestions that helped inform the chapter on animal welfare, and Mark Post graciously read the chapter on lab-grown meat. Also, I want to thank Hod Lipson and Mark Oleynik for chatting with me about 3-D printing and robotics. The folks at BPI graciously hosted me and my students and took us on a tour of their facilities. My fellow agricultural economists, Terry Griffin and Will Masters, helped point me in the right direction when I asked about precision farming and prize-funded research. Frank Yiannas was extraordinarily helpful in educating me about food safety innovations, and Stan Bailey and Kevin Myers graciously spent time talking to me about their companies' efforts to improve food safety.

Although I live in the same small town and work for the same employer as several people I interviewed for this book, I was amazed at how little I knew about the intriguing work of my neighbors and colleagues. It's an honor to live in the same community as such passionate and accomplished people. David Waits and his sons Matt and Mark helped me tell the story of SST and the development of precision farming, and Bill Richardson gave an instructive tutorial on the SST software. Bill Raun and Brett Carver graciously took the time to tell me about their work at Oklahoma State University.

I am grateful to my colleagues and administrators at Oklahoma State University for providing a great place to work and for supporting my research and writing activities. Bill Buckner answered some questions for me about the Noble Foundation and kindly helps support my work with the Oklahoma Council of Public Affairs. The views expressed in this book are my own and may not reflect those of my employer or the people I interviewed or whose work I cited.

I thank my agent, Mel Berger, for sticking with me and finding a home for this book and Karen Wolny, my editor at St. Martin's Press, for taking on the project.

1

Overcoming Nature

There has been a nightmare bred in England of indigestion and spleen among landlords and loom-lords, namely, the dogma that men breed too fast for the powers of the soil; that men multiply in a geometric ratio, whilst corn multiplies only in an arithmetical; and hence that, the more prosperous we are, the faster we approach these frightful limits. . . . Henry Carey of Philadelphia replied: "Not so, Mr. Malthus, but just the opposite of so is the fact." . . . It needs science and great numbers to cultivate the best lands, and in the best manner.

—Ralph Waldo Emerson, 1858[1]

Thomas Jefferson. George Washington Carver. John Harvey Kellogg. Percy Spencer. Who are these guys and what do they have in common? They were food and agricultural entrepreneurs. Their delicious innovations led to new healthy, tasty, convenient, and environmentally friendly comestibles. The creations were *unnaturally* delicious. Unnatural because the foods and practices they fashioned were man-made solutions to natural and man-made problems.

Innovating our way to a brighter food future is as American as apple pie. Benjamin Franklin was an all-around tinkerer. Thomas Jefferson treated his vegetable garden as a laboratory, trying out seeds from across the Old World to carefully select what would work in the new. He even confessed to illegally smuggling rice seeds out of Italy in his coat pocket. The innovating tradition of Eli Whitney's cotton gin and George Washington Carver's peanut creations carried right through to Kellogg's new cereal, Busch's new beer, John Deere's plow, and George Harrison Shull's hybrid corn. Percy Spencer, a radar engineer, created the first microwave oven after finding a melted candy bar in his pocket. TV dinners, Betty Crocker cake mix, Tang, and Lunchables saved time and made home life easier for millions of American men and women. We've inherited a bountiful world of food. One that our ancestors could scarcely have imagined.

La Grand Épicerie is the magnificent food market of Le Bon Marché, the finest Left Bank department store in Paris. Walk in the door and you're greeted by pistachio macaroons, *chouquettes* (pastry puffs covered in scrumptious sugar crystals), beef filets coated with pâté, Roquefort and Chèvre cheese, and wines so expensive they are kept behind lock and key. The sights and smells are enough to tantalize even the least discerning of food palates. My wife's first reaction to the menagerie of temptations that met us on our first grocery-shopping trip while living in Paris? Tears. Not tears of joy, mind you.

The variety and abundance overwhelmed her. What to choose? Will fromage blanc substitute for sour cream? Which brand? What to do with a sea of cheeses of all shapes, sizes, and flavors when one is looking for something simple to top a Triscuit? Of course, we faced the same dilemma back home. Last time I checked, there were more than a hundred different types of bread at our local Walmart. We scarcely notice the abundance because it is so common.

Beneath the tranquil calm of diverse, healthy, affordable food runs an undercurrent of obesity, diabetes, food insecurity, climate change, and environmental degradation—a confluence of forces so powerful that they threaten to upend our very way of life. Alarm bells are sounding amid talk of animal cruelty, unsustainability, corporate farms, and the marketing of junk food.

The problems are real. They are serious. Yet it would be a mistake to think ours was the first generation to have food problems. Or a way out. Carver, Deere, and Kellogg are old news. Fortunately, a new generation of scientists, entrepreneurs, and progressive farmers is carrying on the hunt for unnaturally delicious foods. The problems upon which they've fixed their sights are both new and old.

Before the dawn of the nineteenth century, the eminent scientist Sir William Crooks stood before the British Association for the Advancement of Science and shared these near panic-stricken thoughts:

Civilised nations stand in deadly peril of not having enough to eat. As mouths multiply, food resources dwindle. . . . It is almost certain that within a generation the ever-increasing population of the United States will consume all the wheat grown within its borders, and will be driven to import, and . . . will scramble for the lion's share of the wheat crop of the world.[2]

In expressing what became known as the "wheat problem," Crooks resurrected a then-century-old concern brought to the public's attention by the British cleric and economist Thomas Malthus: If population continued to grow at an exponential rate, there simply wouldn't be enough land and other resources to sustain an ever-hungrier human race. "Misery and vice" was the phrase Malthus used to describe the consequences and cause of the cycles of population growth and privation that he predicted.

Fewer than 1 billion people were living on Earth when Malthus fretted. There were about 600 million more by the time Crooks became concerned. Today more than 7 billion of us inhabit the planet. Despite the impending doom foreshadowed by Malthus and Crooks, La Grand Épicerie has more than we can ever want. And even if wallets are a little thin, one of those inexpensive bread choices from Walmart will surely suit our needs.

How did we avert the mass starvation predicted by the leading intellectuals of the seventeenth and eighteenth centuries? Ralph Waldo Emerson's quote at the beginning of the chapter came from a talk he gave to a group of farmers in 1858. Even then Emerson recognized the key to escaping the Malthusian trap. He said, "We must not paint the farmer in rose-color" but rather look to see that "he is habitually engaged in small economies."

The farmer becomes more productive by planting fences, using underground drainage systems to direct the water, and, even in 1858, using creative fertilizers. Emerson wrote that the farmer "will attend to the roots in his tub, gorge them with food that is good for them. . . . If they have an appetite for potash, or salt, or iron, or ground bones, or even now and then for a dead hog, he will indulge them. They keep the secret well, and never tell on your table whence they drew their sunset complexion or their delicate flavors." Today the key to averting the Malthusian problem is what it was then: innovation.

We tried new things. We tinkered. We invented. We made mistakes. And we tried again. The result is that we now get more than 500 percent more corn and 280 percent more wheat per acre of planted farmland than we did a century ago. Today in the United States we produce 156 percent more food than was the case in the late 1940s despite using 26 percent less land.[3]

Yet public intellectuals remain worried. Lester Brown, founder of the World Watch Institute and the Earth Policy Institute, argued in 1965 that "the food problem emerging in the less-developing regions may be one of the most nearly insoluble problems facing man over the next few decades." In 1974 Brown argued that farmers "can no longer keep up with rising demand; thus the outlook is for chronic scarcities and rising prices." As late as 1997 Brown projected that "food scarcity will be the defining issue of the new era now unfolding."[4] He is by no means alone. In 1968 the Stanford biology professor Paul Ehrlich penned his bestselling book *Population Bomb* with the following prediction as the opening sentences: "The battle to feed all of humanity is over. In the 1970s hundreds of millions of people will starve to death in spite of any crash programs embarked upon now. At this late date nothing can prevent a substantial increase in the world death rate."

It would be easy to pick on Malthus, Ehrlich, and their fellow prognosticators. Making predictions is risky business. Given the information available to them at the time, the impending trends seemed to lead to unmistakable conclusions (and dire forecasts create good publicity, too—Ehrlich appeared on the *Tonight Show* with Johnny Carson more than twenty times). In many ways these folks were right. We *were* headed for doom. We would have witnessed mass starvation, *if* the status quo had prevailed. What made the prophets wrong was how people responded to the challenges they faced. And we should probably be thankful for the soothsayers. Without the doubt and worry they incited, the motivation to change and innovate might have come too late. In Crooks's case he actually pointed the way to scientifically working around one of the greatest limiting factors in agriculture: extracting nitrogen fertilizer from thin air.

Even a brief glance at today's popular writings reveals that pessimism about food and agriculture abounds. If anything, the apocalyptic forecasts have increased and expanded into ever-new areas of concern. Distrust of the food system has become the status quo. At least another billion people are likely to join the human ranks in the next thirty years. Researchers for the United Nations project an 80 percent chance that the world's population will increase by 70 percent, to 12.3 billion people, by the year 2100.[5] As late as 2014 Ehrlich and a coauthor argued there was only "about a 10 percent chance of avoiding a collapse of civilization."[6]

But population growth is not the only concern. Today's food problems are complex and multifaceted. Worries about the environmental impacts of food production persist. Although we get much more food from our land than in Malthus's time, we also use more fossil fuels and have created problems like dead zones in lakes and waterways caused by excessive fertilizer use. For many people in the developed world, there are problems of *over*abundance: obesity and growing rates of diabetes. Marion Nestle, a nutrition professor at New York University, argued that the costs of obesity and diet-related diseases will be "astronomical," and James Hill, director of the Center for Human Nutrition at the University of Colorado Health Sciences Center, argues that diabetes alone "will break the bank of our healthcare system."[7] The bestselling author Michael Pollan summed up the prevailing view he helped cement: "Americans have a national eating disorder."[8]

We have problems. But we've had them before. Many would argue that the state of food in the United States is not a happy story. However, below the surface of the food problems are churn, change, and innovation. Much of what is happening is imperceptible to the average American just trying to put food on the table for the family. People have a tendency to focus on such headlines as "Obesity Is Rising Out of Control!" without seeing the progress being made: the trend is that waistlines no longer are expanding at the pace they once were.[9] The Green Revolution sparked by Norman Borlaug, who was awarded the Nobel Peace Prize in 1970, lifted millions of people out of desperate conditions in the 1960s and 1970s by introducing hybrid seed technologies and synthetic fertilizers in places like India, Mexico, and Pakistan. There remain places on Earth that could still benefit from applying the concepts and technologies introduced by Borlaug, but, like the Apple IIe, science and technology must adapt if they are to remain relevant.[10]

Two narratives currently dominate popular thinking about the future of food. The one with the cultural cachet is the so-called food movement—a movement that seeks a retrogressive "return to nature." The food movement has issued a call to eat slower, more natural, organic, local food. Those are all good things in their own right. The trouble is that most of us aren't willing to pay (and many are not able to pay) what it costs to produce food that way. And most farmers aren't willing to give up their modern conveniences without sufficient compensation.

Faced with this impasse, many leaders of the food movement say the answer is to subsidize the food systems they like, tax and guilt-trip the people eating the foods they don't, and regulate the undesirable foods and farming practices out of existence. I exposed the ineptitude and unintended consequences of many of these ideas in *The Food Police*. The food movement is a compassionate, romantic cause, but only the most credulous will believe that its most visible policy activities—advocating soda taxes, mandatory labels for genetically modified organisms (GMOs), subsidies for fruit and vegetable farmers, and support for farmers' markets—will have a substantive effect on any of our most pressing food challenges.

The counternarrative is made by the heirs to the Malthusian concern. They say the answer to feeding a rising world population is to produce more, whatever the cost to health or the environment might be. Quantity over quality. Efficiency trumps humanity. Subsidize largess. Consumer preferences be damned.

Is there a way to have the best of both worlds? More affordable *and* healthy food for our families? Environmentally friendly *and* convenient food? Surely you have been warned about politicians who offer a free lunch. We simply cannot have lower taxes *and* build more roads. There are trade-offs. Yet there is a sense in which scientific and technological development offer at least an inexpensive lunch. As the author and MIT researcher Andrew McAfee recently put it, "The old joke among economists is that technology progress is the only free lunch we believe in."[11] If entrepreneurs develop a new way to make concrete or if managers find more effective use of labor, it might very well be possible to have more roads *and* lower taxes.

To get these (almost) free lunches, we will have to make some trade-offs and be willing to do things differently than in the past. That includes giving up not only current practices but also the way our grandparents and great-grandparents farmed: it was drudgery, and it wasn't sustainable. And we may need to be willing to change the way we think about our food problems.

A couple years ago I gave a talk at a small liberal arts college. Before my talk I had dinner with about a dozen bright students who were well versed in the popular writings about food and agriculture. Smart and motivated, with an elite education, they had futures that shimmered with promise and hope. And yet, when the topic turned to food and agriculture, pessimism abounded. The students peppered me with questions and expressed fear and skepticism related to unfair farm labor practices, genetically engineered crops, pesticides, soil runoff, and animal abuse.

They were doing what they could to address these problems in their own way. The students earned course credit by working on a small farm owned by the college, part of the college's sustainability curriculum. The students (with the help of faculty advisers and administrators) encouraged the school cafeterias to spend about 40 percent of the food budget on local products. Some students volunteered their time to work on small local farms.

A couple students were a bit startled to hear that I intended to spend about half my talk that evening discussing the research that directly challenged the idea they had been taught to cherish: that local foods are inherently better for the environment, health, and the economy.[12] I received a warmer reception when I mentioned that I planned to talk about another way we can solve many of the food problems they cared so passionately about. After I described how scientific breakthroughs and technological development in agriculture can spare sensitive lands and bring down the price of food for the poor, one earnest young woman caught on and remarked that what we really need are people working on soil and plant sciences and on environmental impacts of animal production. I was happy to report that there are.

I suspect so much of the negativity that surrounds our food discussions stems from a sort of hopelessness that comes from the inability to see how our problems will be solved. Compared to my fellow diners, I'm much more optimistic about the future of food, in part because I have the great pleasure to work at an agricultural college where I see firsthand good, intelligent people applying their scientific craft in the pursuit of better farming and better food. I look out at the sea of plant pathologists, soil chemists, ruminant nutritionists, agricultural economists, food engineers, ecologists, range scientists, microbiologists, and many others who are hard at work at universities and research centers throughout the world. I routinely encounter food and agricultural entrepreneurs asking for help in raising capital for their new ventures; jaw-dropping developments in the labs of seed, biotechnology, and food companies; as well as everyday farmers who are eager to use the latest technology.

This is the story of the innovators and innovations shaping the future of food. I'll introduce you to David Waits, the farmer-turned-entrepreneur whose software is now being used on more than 100 million acres in twenty-three countries to help farmers increase yields and reduce nutrient runoff. You'll meet Tom Silva, who helped his employer build a new hen-housing system that improves animal welfare at an affordable price. Mark Post is a scientist whose work may lead us away from eating animal products altogether. He's growing meat in his lab. Without the cow. I'll take you behind the scenes of a student competition at which Sarah Ritz and Aaron Cohen coaxed bacteria to signal when olive oil is stale and Paul Tse and Marco So engineered a probiotic to fight obesity. I'll take you to South Dakota, where Eldon Roth created a new way to fight food waste. You'll learn about work by my former student Abdul Naico and the German scientist Ingo Potrykus that aims to fight malnutrition in the developing world with nutrient-enhanced rice and sweet potatoes. My plant science colleagues at Oklahoma State University reveal how they're helping wheat farmers sustainably grow more with less. And the engineering professor Hod Lipson discusses how to get fresh, tasty, 3-D printed food at the touch of a button, perhaps even delivered to us by Mark Oleynik's robotic chef.

Some of this might seem a bit scary or even unappetizing. But the same could once be said of the refrigerator and the microwave. And broccoli. And kale. These are *un*natural *human* inventions. Before people had refrigeration, they ate a lot more canned and salted meat and had to have milk delivered every day. Some food elitists bemoan the microwave, but come on! Would you really give up yours? Broccoli, cauliflower, cabbage, brussels sprouts, and kale didn't exist before humans came along. All these veggies are descendants of the same plant, and they originated through artificial selection. Nothing seems more natural than Irish potatoes or Italian tomatoes, but these plants arrived in Europe only after Columbus sailed the ocean blue.

There is a tendency to want natural food—to eschew the foods that we humans have tinkered with. I've had more than one person tell me they just want to "eat the food that God gave us." The historical reality is that we've been altering our food and innovating new diets since the beginning. Indeed, one thing that separates us from other animals is that we invented the technology of cooking. Rachel Laudan, speaking of our ancestors in her book *Cuisine and Empire,* writes:

> Before the first empires, indeed long before farming, [our ancestors] had passed the point of no return, where they could no longer thrive on raw foods. They had become the

animals that cooked. Cooking softened food so that humans no longer had to spend five hours a day chewing, as their chimpanzee relatives did. It made it more digestible, increasing the energy humans could extract from a given amount of food and diverting more of that energy to the brain. Brains grew and guts shrank. Cooking created mouthwatering new tastes and pleasing new textures. . . . It became possible to detoxify many poisonous plants and soften others that had been too hard to chew, so that humans could digest an increased number of plant species. . . . Ways of treating flesh and plants so they did not rot permitted the storage of food for the lean times of hard winters or dry seasons.[13]

We've been inventing and adapting all along. And we're better for it.

Organic and local foods get the headlines. Farm subsidies, soda taxes, GMO labels, and bans on fast-food advertising are political lightning rods that are sure to get your friends riled up. These things captivate our attention because they give us a way to feel like we're "doing something" by paying a premium or fighting a political battle. But good intentions don't always produce good outcomes.

One of the problems some people may have with relying on technological innovation to address our food problems is that it seems to leave nothing for us to do. How can the passionate food activist, sociologist, or chef use science and technological innovation to produce the food system they desire? The first step is to recognize the innovation that is under way. Some technology adopted by farmers is a bit frightening because the public is so unaware. Fewer than 2 percent of Americans work on a farm. Thus many find it difficult to understand why farmers adopt certain practices or technologies or what problems they may be solving when they do. Getting their perspectives can help shed some light. Moreover, as I will discuss, food and agriculture innovation doesn't come only from Monsanto, Cargill, and McDonald's. It comes from students, nonprofit scientists, university professors, and struggling entrepreneurs. Fostering an environment that is hostile to innovation and growth in food and agriculture not only thwarts the plans of Big Food but also makes it harder for scientists to get their innovations to market.

If I accomplish nothing else with this book, I hope a few young people might see a new way to effect food change. Yes, take classes in food journalism and environmental sustainability. But don't forget mathematics, biology, geography, engineering, and genetics. Ironically, the greatest outcomes from study of the natural sciences may well be all the unnatural things we learned to create: planes, cars, iPhones, air conditioning, and vaccines. We may romanticize the past, but most of us would not wish to be born in 1800. Changes in automotive, medical, computing, communications, *and* agricultural technology deserve the credit. Life—particularly in the realm of eating—*is* substantially better today than it was in our great grandparents' time. And, if history is our guide, it will become better still. Let me tell you how.

2

The Price of Happy Hens

When I was in college, one of my food science professors would often tell us that eggs were a near-perfect food. They are a complete source of protein, containing all the essential amino acids that our body can't make on its own. Fear of cholesterol caused a significant reduction in per-capita egg consumption throughout the 1960s, 1970s, and 1980s, but the latest report from the Dietary Guidelines Advisory Committee (the group responsible for creating the food pyramid and MyPlate) suggests those fears were unfounded and now says that "cholesterol is not a nutrient of concern."[1]

Not only are eggs nutritious, but it's hard to imagine how we could make some foods without them. Obviously, dishes like omelets and deviled eggs would have to go, but eggs are also the crucial ingredient in mayonnaise and salad dressing: they bind the water and vinegar to the oil. Hollandaise and bearnaise sauces, spaghetti carbonara, quiche, cake, and pie all rely on eggs as a key ingredient.

New companies like Hampton Creek, the maker of Just Mayo, are trying to replace the egg with a specially engineered mix of yellow pea proteins.[2] It's too early to tell how successful Just Mayo might be, but previous efforts to sell egg substitutes haven't gained much traction. For now it appears our eggs will come from chickens. A lot of chickens. In 2014 the United States was home to more than 300 million hens, who laid 86.9 billion eggs for our dinner table.[3] Add it all up and the average American ate about 260 eggs last year—an amount roughly equal to what a single chicken lays in a year.[4]

We tend to have romantic views about the lives of the chickens laying all these eggs. That is, if we think about it at all. Most consumers have little idea of how eggs are produced, and the public tends to have an idealized view of agricultural production. For example, a colleague and I asked several hundred consumers what percentages of eggs in the United States are laid in cage versus cage-free systems. On average the respondents thought only 37 percent of eggs come from cage systems. The reality is that more than 90 percent do.[5] Moreover, after we gave consumers unbiased information about different types of animal housing conditions on U.S. farms, more than 70 percent of consumers reported having greater concerns about the well-being of farm animals.

It seems the more people learn about the living conditions of farm animals—particularly egg-laying hens—the more concerned they become. Sometimes these concerns translate into political action.

In November 2008 Californians went to the polls to vote for a new president. Sixty-three percent of those voters also chose to give egg-laying chickens in the state a new home. As in the rest of the country, hens in California lived in so-called battery cages so small that the hens cannot completely stretch their wings. Hens living in these housing systems typically have about sixty-seven square inches of space—which is 28 percent less space than a typical 8.5-by-11-inch sheet of paper.

I suspect most people would applaud the mandated space increase as self-evidently positive. As it turns out, the story is more complex.

One problem is that Californians used the political process to ban a practice that they routinely embrace in the marketplace. Even before the vote, virtually every major grocery store sold cage-free and organic eggs. Yet fewer than 10 percent of Californians were willing or able to pay the extra cost for eggs produced that way (cage-free and organic eggs were 20 to 80 percent more expensive than eggs produced by caged hens in California).[6] The result was that voters told California egg farmers to adopt a more expensive production system without sufficient compensation from consumers to cover the costs.

This vote-buy paradox is no academic quandary for Tom Silva, vice present at JS West, a 106-year-old company owned by a family and its employees and based in Modesto, California.[7] Silva is responsible for the production and well-being of about 1.5 million hens housed at farms owned by the company.[8] Silva could choose to simply give hens more space, but to do so would entail higher costs. Higher costs would curtail egg consumption, particularly among lower-income consumers. Did I mention the nutritional value of eggs?

In fact, the new California law forces Silva and all other egg farmers in the state to incur higher costs if they want to stay in business, even though they are unlikely to reap any additional money from the market. Silva and other egg producers in California soon realized that the new law would put them at a cost disadvantage in relation to egg producers in Arizona, Idaho, and Iowa.

When the voters approved the proposition, nothing prevented California grocery stores from simply importing the less expensive battery-cage eggs from other states that had no such law.

Not surprisingly, California egg farmers barred from producing battery-cage eggs soon convinced the state legislature to outlaw grocery stores from selling these sorts of eggs as well. Whether the sales ban can withstand legal challenges that argue it violates the Commerce Clause of the Constitution remains to be seen. What is clear is that California farmers are struggling to figure out exactly how to house their chickens. And because California is such a large state (and imports a large number of eggs from other states), the laws there are affecting egg farmers throughout the country. In fact, the state legislatures of Oregon and Washington subsequently passed their own laws regarding hen housing.

The answer to the conundrum might seem simple. Why not simply adopt the cage-free production systems that provide high-end eggs to places like Whole Foods? When we see cage-free eggs in the store, with their fancier packaging, brown shell, and invariably higher price tag, it's easy to imagine hens frolicking in a scene from a children's storybook. Old MacDonald had a farm, and it was a cage-free farm. But we don't live in a storybook world.

Typical cage-free systems (often called barn or aviary systems) provide hens with much more space than do the cage systems. The barns allow the birds to exhibit natural behaviors like scratching and dust bathing, and they provide nesting areas for laying eggs. But they are far from the paradise many people envision. As Silva put it, "Cage-free isn't what most people think it is."

The barns or aviaries are often chaotic, dusty, and smelly. Mortality rates for cage-free hens can be twice as high as those for hens in cages. So even though the hens have more amenities and freedom than in the battery-cage system, they die at a much higher rate. Some of that is a result of more fighting (the phrase "pecking order" is not some abstraction but a reality in hen houses). Higher death rates are also partially attributable to the different breeds of chickens typically used in cage-free systems, Rhode Island Reds, which lay brown eggs, whereas White Leghorns, which lay white eggs, are typically used in cage systems. But the higher mortality in the cage-free systems can also be partially attributed to conditions that are less sanitary. Air quality is particularly bad, as are particulate matter emissions. This is bad news for the birds, and many employees also don't like it. I've talked to large-scale egg farmers who have both cage and cage-free systems, and most prefer the cage. In addition, cage-free systems have higher carbon footprints and produce eggs that are 30 to 40 percent more expensive than eggs from cage systems.[9]

If all that's a bit depressing, maybe all of us should stop by Williams-Sonoma the next time we're at the mall. Last time I checked the catalog, you could get a backyard cage big enough for about six hens that will run you $300 to $1,500.[10] No doubt backyard chickens have become fashionable. My neighbors have some. Their eggs are tasty, and the hens are fun to watch. The girls look like they're enjoying themselves as they roam around the yard. Except when the temperature drops below freezing. Or when the mercury tops out at more than 100 degrees for months on end. Or when the hawks, coyotes, foxes, or neighborhood dogs come prowling.

There's a reason farmers started bringing their hens indoors decades ago. It wasn't because they were evil "factory farmers" but because they could provide a safer and more stable environment for the hens. It also allowed the farmers to more closely monitor the hens' diets. Eating bugs and grass and dirt is all fine and well—and these varied feedstuffs can produce tastier eggs—but such diets can also convey parasites and disease. One of the big concerns is the spread of avian influenza (or bird flu) through backyard chickens.[11] In 2015 millions of turkeys and chickens died or were euthanized because their flocks were infected with avian influenza thought to have been spread by wild birds.[12] I'm not necessarily trying to dissuade backyard chicken owners, but land, space, time, knowledge, and income constraints mean most Americans aren't going to have backyard hen houses. Trying to imagine how each and every American could get their annual consumption of 260 affordable eggs from backyard hens boggles the mind, and the very notion stretches credulity.

Is there a middle ground? An innovative solution?

Silva thinks so. He helped JS West become the first egg producer in the United States to adopt a new kind of housing system—an enriched, or colony, cage system (sometimes also referred to as a furnished or modified cage). Silva realized the new California laws would dictate change, but simply adopting the cage-free aviaries didn't seem like the right approach for many of the reasons I've discussed. Silva and JS West looked east toward Europe. More than a decade before Californians decided to outlaw battery cages, several European countries had done the same. Aware of the pushback against the battery cage system, animal welfare researchers in Europe had been working for years on alternatives that could combine the advantages of the cage system with the advantages of the cage-free systems while avoiding some of the worst drawbacks of each.[13] Jon Bareham, a British animal welfare researcher, proposed a "get-away cage" in the late 1970s.[14] The idea was to make a cage that had perches and nests on different levels where hens could get away from attacks by bullying birds and nest in a secluded place. His ideas were further explored and studied by Dutch and German scientists.[15] The development of enriched cages that included furnishings like perches, scratching areas, and nesting areas ultimately was brought about by Mike Appleby and others at the Poultry Research Centre in Edinburgh, Scotland.[16]

Silva traveled to Europe to take a look at the enriched cage system. They are conspicuously bigger and more spacious than battery cages: the enriched cages provide 73 percent more space per hen. Although the set-up differs from farm to farm, a typical enriched cage provides about 15 percent more space than a king-sized mattress.[17] The much larger cage is home to sixty hens—providing about 116 square inches per hen instead of 67.

In a typical battery-cage system, hens lay their eggs on a slanted wire floor (the eggs roll to the edge of the cage, where they fall onto a conveyer belt). It is the same wire floor on which the hens routinely stand. In contrast, hens in enriched cages may be on perches mounted over a wire floor or in a staging area with a mat for scratching, or they may be in a secure nesting area when they lay their eggs. The hens must like the nests because 95 percent of the time they enter the nesting area to lay their eggs, which gently roll onto a conveyer belt before being whisked away into an adjacent sorting and washing room.[18]

Unlike the barren environment in the battery cages, the enriched colony cages have the mat area that allows the hens to exercise their natural urge to scratch. Also available are perches that allow the hens to get up off the wire floor. In addition to the nests, the perches are a popular sleeping area for the hens. Running underneath the colony cage is a conveyor belt that removes the manure and keeps it away from the birds.

The enriched colony cages aren't perfect, and some animal advocacy groups think they don't go far enough. But they're an innovative compromise. Indeed, two groups that are often foes—the Humane Society of the United States (HSUS) and the United Egg Producers—worked together for a time to try to make the enriched cage system a national standard.[19]

Silva is proud of the new barns he helped build at JS West, and the company plans to build more. The company is so proud, in fact, that its website has videos of the operation and streaming web cameras in its barns so that customers can check in at any time to see how hens are being treated.[20] In an era when consumers demand greater transparency, JS West—through the power of the Internet—has provided it. This is more transparency than you'd get from a vendor at a farmers' market. But transparency isn't always appreciated. Silva says he sometimes receives angry or negative comments about the live camera feed (some of it likely from coordinated efforts by activist groups). Some of the outrage reflects our agrarian idealism, while other protesters are simply letting the perfect be the enemy of the good.

None of that is to say we can't do better. The challenge is that the choice between free-range, cage-free, and battery-cage housing entails tough trade-offs between egg affordability, worker health, hen mortality, and hens' freedom of movement. The way around these trade-offs is innovation. The enriched colony cages give the hens more room and amenities without the spike in mortality and dust that tends to come with the cage-free systems—all at a cost only somewhat higher than what we expect to pay for eggs.

More innovation is possible. For example, one of the downsides of the enriched colony cage system is that hens don't have access to the outdoors. My research suggests that we humans tend to think outdoor access is an important component of animal welfare.[21] Yet we're probably suffering a bit from anthropomorphism. It's not that hens wouldn't like to go outside from time to time, only that outdoor access ranks relatively low among the factors that influence hen welfare. One scientific study, for example, ranked twenty-five different issues that could affect hen well-being, and outdoor access, or the ability to range free, ranked only nineteenth. Having nests is three times more important than access to the outdoors with ample shelter, and having a dust-bathing area is twice as important in boosting hen welfare.[22] Thus, if we want to provide hens with outdoor access, we need to do it in a way that actually improves hen welfare in an affordable and responsible manner.

One group of Dutch researchers has been working to create just such a system—the Roundel (the eggs are sold in a circular, biodegradable carton under the name Rondeel).[23] The name comes from the shape of the circular barn devised by researchers at Wageningen University, one of the top agricultural universities in the world. The circular barn is divided into different slices like a pie, with each slice offering a different amenity for the hen. One slice is for nesting, another for perching, and another for indoor foraging and dust bathing. The center of the pie houses the egg collection, cleaning, and storage operation. One slice of the pie is reserved for trucks bringing in feed and taking away eggs. Surrounding the circular barn (minus the slice reserved for transportation) are fenced-off areas that provide outdoor access and that can be open or closed depending on weather conditions or threats of avian influenza.

The Roundel is the Ritz Carlton of hen living. Hens have virtually all the freedoms and amenities they'd want from the wild but with ample feed and without any of the dangers from predators or hardships from adverse weather. The Roundel also comes with a luxury hotel price. When I checked the prices of eggs in a large Amsterdam supermarket, a round-pack of seven eggs from Rondeel cost $2.23, or about $0.31 per egg. There were much less expensive eggs for sale at the store, such as the plain carton of white eggs selling for $0.14 per egg, but there were also a couple more expensive organic (or "bio," as the Europeans call it) egg cartons with a sticker price of more than $0.34 per egg.[24] There isn't yet much research on the environmental and animal welfare impacts of the Roundel system, and it remains unclear whether further innovation can bring the costs down to a more affordable level. Yet these sorts of science-led innovations have the potential to improve the lives not only of the hens but of us as well.

* * *

There also are some exciting innovations in hog and dairy cattle housing. However, new housing systems or more transparency are unlikely to address all the concerns about animal welfare. Even though many people appreciate the changes being made at places like JS West, Silva lacks a way to recoup his company's investment in the improvements in hen well-being because many people aren't willing to pay a sufficient amount to offset the extra costs. After three years of trying to separately brand and market eggs produced in the enriched cage system, JS West gave up. Silva said the least expensive eggs sell the best, and it was tough to justify the added marketing costs, especially if, after the change in law in California, the enriched-cage eggs become the lowest-common denominator commodity egg.

The current market environment poses an altogether different challenge for animal advocates. Some of the people most passionate about farm animal living conditions don't eat meat or eggs. A vegan who is troubled by farm production practices can't switch from cage to cage-free eggs. Nor can they eat less meat. They aren't buying any to begin with. How can the vegetarian, vegan, or infrequent carnivore encourage the changes they want in the food system? It is not surprising that these folks turn to protest, activism, the courts, and the ballot box in an effort to improve farm animal welfare.

Usually when we want something, we search the Internet or head to the supermarket to find someone willing to sell it to us. We may not always make a purchase if the price is high, but we know we can generally turn to the marketplace to find the things we want. The profit incentives embedded in the market economy prompt entrepreneurs to find, create, and manufacture the things we want to buy.

The crux of the problem is that it is not possible to directly buy animal welfare. You can buy cage-free eggs, but—at least at present—you can't directly buy happier hens. What we need, then, is a market that allows us to directly buy the thing we want, improved animal well-being, whether we want eggs or milk or bacon—or not.

If all this seems a bit far-fetched, perhaps it might be useful to consider an analogous market: the market for pollution. When coal plants make electricity, they also produce pollution, an output not traded in the market. Back in the 1960s a few economists proposed "pollution trading" as a more effective way to deal with environmental problems than command-and-control policies that dictate which technology can be used or how much pollution can be emitted. Many of these ideas came to fruition with the passage of the 1990 Clean Air Act, which established the first large-scale tradable emissions permits to curb the sulfur emissions responsible for acid rain. The program set a cap on the emissions that a power company could emit, but the policy allowed those companies wishing to generate more pollution than the cap to buy "allowances" from those companies that produced less pollution. In so doing, the policy provides a financial incentive for power companies to adopt the lowest-cost methods of curbing pollution. The more energy efficient the company becomes, the fewer allowances it has to buy. If the power plants become efficient enough to fall under the cap, the company could even profit by selling allowances.

The emissions-trading markets have been heralded as a great success. For example, a group of MIT economists write,

> Not only did [the market trading program] more than achieve the [sulfur] emissions goal . . . it did so on time, without extensive litigation, and at a cost lower than had been projected. . . . We have learned that large-scale tradable permit programs can work roughly as the textbooks describe; that is, they can both guarantee emissions reductions and allow profit-seeking emitters to reduce total compliance costs.[25]

Similarly, the Harvard economist Robert Stavins has declared, "Market-based instruments for environmental protection—and, in particular, tradable permit systems—now enjoy proven successes in reducing pollution at low cost."[26]

It is possible to imagine a similar approach to animal welfare—a mandated minimal level of animal well-being (farms that
have lower levels of animal welfare must buy allowances from
other farms that provide higher levels of animal welfare), but
purely voluntary schemes are also possible. One example is the
active voluntary markets for carbon emissions aimed at curbing
climate change.

At present no federal regulations force U.S. firms to limit
carbon emissions and yet markets like the now-defunct Chicago
Climate Exchange and the still-active European Climate Exchange allowed companies and municipalities to offset their carbon emissions by buying credits produced by farmers and forest
owners willing to make changes to store and sequester carbon.

Some cities and municipalities participated in the exchanges
in response to voter pressure to reduce carbon emissions. Many
companies did the same as a part of green and sustainability initiatives promoted by shareholders and as a part of public relations
campaigns. And many individuals purchased carbon offsets to
salve their own conscience. Although these exchanges were voluntary, the world market for carbon trading involved more than
$60 billion in trades in 2006.[27]

Today, if you want to counter the carbon impacts you create from flying, you too can buy carbon offsets. Offsetting the
carbon impacts of one 6,000-mile flight will cost you a little
more than $11 at carbonfund.org. The annual carbon impacts
of eating beef can be offset for less than $10.[28] The organization
uses the money to fund projects like planting trees that sequester
carbon.

What does pollution trading have to do with animal welfare? Just as firms create carbon when making cars or electricity, farmers create animal welfare (both good and bad) when they make eggs and milk and bacon. If we can measure the level of animal welfare a farm provides, the farmer can receive a certain amount of credits—or animal well-being units (AWBUs), as I called them in a paper for the journal *Agriculture and Human Values*, where I introduced the topic.[29]

Once a farmer earns credits, the farmer can sell the credits to anyone wishing to improve animal welfare. In such a system people passionate about animal well-being have a direct and tangible means to get what they want. Such a system would achieve an overall level of animal well-being that balances the costs of providing higher levels of care with people's demand for it. The price of AWBUs would be determined by the interaction of buyers and sellers, and it would no doubt fluctuate over time in response to changes in demand for animal welfare and in response to changes in the costs of providing it.

Creating AWBUs may seem fine in theory, but is it possible to actually quantify animal well-being so that it could be tradable? Difficulties are involved in such a calculation, but they are not insurmountable, no more so than, for example, measuring the carbon impacts of different types of food production. A variety of approaches are available to determine the overall well-being of animals in housing conditions. These approaches range from expert opinion to simple checklists to formal models that rank various measures on the farm, with more weight given to those measures that scientific studies deem more important to animal well-being. Numerous animal welfare auditing and certification programs are already in use, which suggests that companies and third-party certifiers are already quantifying the concept of animal welfare. It is true that people may differ in their subjective beliefs about the effect of certain factors on animal well-being (for example, how important is providing hens with nests versus scratching space?), but this need not hinder the creation of a market for animal well-being any more than trading in the market for computers, cars, and food depends on subjective beliefs about the merits of brand names, *Consumer Reports* ratings, the recommendations of friends on Facebook, and so on. If potential traders don't like the way animal welfare is calculated in a particular market, they don't have to buy AWBUs, but this need not prohibit individuals or organizations from creating a market around a particular concept of farm animal welfare.

I prefer the use of mathematical models that look at specific, individual issues on a farm—the amount of space, air quality levels, animal health, and so on—and aggregate them to create an overall score related to the well-being of that farm's animals.[30] For example, at the high end, farms where animals have ample space, food, and access to amenities and are free from stress and disease might get a score of one hundred, whereas farms with high mortality and high injury rates, and where animals are housed in cramped and austere conditions, might get a score of zero. Farms with intermediate conditions would receive intermediate scores. These scores could be used, along with the number of animals on the farm, to assign a certain number of AWBUs to the farmer.

Once in possession of the AWBUs, the farmer can bring them to market and offer to sell them to any willing buyer.

The potential buyers of AWBUs are numerous and diverse. People donate billions of dollars to charities each year, and it is plausible that donors might use some of this money, either directly or through the charities to which they donate, to purchase AWBUs. Take, for example, the Humane Society of the United States. The HSUS has more than 2.2 million Facebook fans and reported donations in excess of $186 million in 2014.[31] Some of this money was spent in legislative battles and in other public relations campaigns that, at best, have an uncertain and indirect effect on animal well-being. Supporters of the 2008 California ballot measure that outlawed the smallest battery cages spent $5.2 million to convince voters to approve it.[32] Some of the HSUS's $186 million budget, or the $5.2 million spent by supporters of the California iniative, could have been spent on AWBUs, which would have had a certain and direct effect on animal well-being.

The HSUS is only one of many animal advocacy organizations that spend millions each year attempting to improve the well-being of farm animals. However, these organizations are not the only potential buyers of AWBUs. Companies like McDonald's and Walmart, for example, might want to make public statements about their commitments to animal well-being by buying AWBUs at a volume proportionate to their use of animal products. And, of course, any individual might buy AWBUs if she wishes to "offset" the impacts of egg purchases. Even if a consumer doesn't buy eggs, she may want to buy AWBUs to improve hen well-being.

Farm organizations have little reason to fight the creation of a voluntary market for AWBUs. The ability to profit is all the incentive a farmer would need to participate. For example, many farmers have completed costly processes to qualify for certification programs, adopted organic practices, or have stopped using growth promoters in animal production for no other reason than that some consumers are willing to pay for these sorts of foods. If creating more AWBUs yields more profit, you can bet some progressive farmers will sign up.

I've presented these ideas to both animal production groups and members of animal advocacy organizations. By and large, farmers and ranchers have liked the idea, especially if it's voluntary. After all, if they don't like the program, they don't have to participate. I've received more pushback from the philosophers and ethicists who advocate for sweeping changes in the way we treat animals. They say things like, "Well-being is infinitely valuable because all animals deserve humane treatment" or "We need to recognize that there are some things that money can't buy and other things that money can buy but shouldn't." But some of this reflects an all-or-nothing mentality. We are unlikely to reach utopia in the near future, but we can make incremental improvements in animal well-being.

The reality is that the vast majority of Americans eat meat, dairy, and eggs. Current levels of animal welfare are already determined by a market. The meat market. Thus the question isn't whether markets should dictate farm animal well-being but rather what kind of market will dictate farm animal well-being. If we don't like how our current meat, egg, and dairy markets have incentivized certain production practices that reduce animal welfare, we can harness that same power to create new, innovative markets that do just the opposite.

3

Hewlett Packard with a Side of Fries

For most of human history, if you wanted a copy of something, you had to make it yourself—by hand. Or pay someone else with more expertise to do it. The printing press with movable type was invented less than six hundred years ago. Before that, most books, particularly religious texts, were replicated by scribes who laboriously copied them letter for letter from older documents or dictation. Sometimes the scribes would leave their own feelings about their task in the margins of medieval documents with notes like "Oh, my hand," "Thank God, it will soon be dark," or "Now I've written the whole thing: for Christ's sake give me a drink."[1]

Fortunately, printing has become much easier. Growing up with two schoolteacher parents, I can vividly remember the noxious smell of the ditto machine whose roller seemed to leave purple ink stains on everyone who walked within eyesight. That those ditto machines now seem so ancient speaks volumes about how much we've progressed just since the 1970s. Even as late as the 1950s the idea of easily making a precise, stain-free copy at the push of a button probably would have been regarded as fanciful.

With the advent of 3-D printing, replication no longer is limited to length and width. Nor is 3-D printing limited to making plastic figurines. One day soon you may be able to satisfy your hankering for a fresh homemade pizza simply by pushing a button on your smartphone.

For about a thousand bucks, it is already possible to buy a kit that can print food, according to Hod Lipson, a robotics and engineering professor at Columbia University; he heads up a project that makes Fab@home, a 3-D printer.[2] It isn't exactly the food replicator used by Captain Kirk, but 3-D food printing and robotic chefs are moving us a few small steps in that direction.

Lipson originally went to school to study robotics and engineering. As a student, he became interested in 3-D printing for making robot parts. In those days, in the early 1990s, 3-D printing was in its infancy and consisted mainly of making plastics. Lipson was interested in a more flexible printing machine—one that could print in materials beyond plastic. Altering the 3-D printer to make wires, batteries, or gels would broaden its usefulness not only for robotics but also for biomedical applications.

Lipson's interest in food printing came accidentally. As a young professor at Cornell University, he faced the challenge of printing figures with "overhang." This is an old problem in 3-D printing. As Lipson told me, it's no problem to print a pyramid, for example, because the upper layers can rest on the larger base below it. But if you wanted to print an inverted pyramid, it would become unstable because gravity would make the unsupported upper layers collapse on the smaller layers before the creation solidified. Lipson had an idea that might ameliorate the overhang problem. He'd already been working on printing with unusual materials, and he looked for a material that could be printed temporarily as support for overhangs and then be washed away.

The answer was frosting. His machine printed frosting as support for other printed material; once the creation was complete, the frosting was washed off. Before long Lipson noticed that his students weren't just washing away the frosting. They were eating it. The idea of printing food for the sake of food was born.

A high school student had a similar idea at about the same time. By the mid-2000s there were commercial 3-D printing machines, but none for the sorts of unusual materials that interested Lipson. He had to create his own machine, and he and his collaborators decided to open source some of the work. They created the first open-source 3-D printing machine in the United States. They listed online the materials needed to create the machine and the software required to run it. The so-called Fab@home machine is the result.

A Kentucky teenager, Noy Schaal, decided to create a Fab@home in 2006. With help from her father, she altered the original design to include a heating element and loaded the machine with chocolate. Before long she was printing chocolates in the shape of her home state. Not only did she win a science fair, but she probably created the first 3-D–printed food.[3]

Once the seed of 3-D food printing had been planted, Lipson worked to make it grow. He talked to chefs and the folks in the culinary program at Cornell. They weren't interested. The chocolatiers, in fact, were appalled, even though a high school student had already demonstrated that printed chocolate was possible. Apparently the lure of hand-crafted naturalness so consumed the chocolatiers that they at first could not see the applicability of 3-D printing. Never mind that French and Swiss chocolates are not natural European delicacies. Cacao made it to Europe as a part of the Columbian exchange.

Food printing finally was embraced by confectioners. The candy makers were accustomed to experimentation, and 3-D printing was another tool they could use to create new treats. Since that initial introduction, 3-D confection printing has blossomed. A company called 3D Systems has a commercial model available—the ChefJet—that prints in sugar. The machine uses an inkjet printer head that sprays water onto sugar, creating crystals. The machine turns out beautiful creations. Some are amazing works of art that look like intricately colored delicacies picked off a coral reef. Other, smaller, bite-sized candies can be printed in myriad colors and patterns. Some of the most unusual look like small jungle gyms. The company has partnered with the Culinary Institute of America to bring the idea of 3-D food printing to the next generation of tastemakers.

Other commercial machines print three-dimensionally in frosting or chocolate (and one model, the PancakeBot, focuses exclusively on printing and cooking pancakes). One of the simplest and least expensive commercial models is made by 3D Ventures, and it seems to be used mainly for cake decoration. More complicated models, like the CocoJet by 3D Systems, produced in collaboration with Hershey, create personalized chocolates. You can print latticed hollow Hershey's kisses, chocolate roses, your name, or virtually any other shape imaginable. The machine works by printing thin layer upon thin layer of chocolate.

Lipson said the big food companies initially were not interested. Their executives would ask him, "How many cookies can be printed per minute?" When he said, "About one-third," the conversation was over. The process remains slow enough that it cannot compete with the line speeds used in cookie manufacturing plants. Yet 3-D printing machines can do things the big industrial machines can't. For one, 3-D printing can readily combine ingredients and mixes in unusual ways—for example, printing names or logos in alternating colors on the inside of a cookie or cake. But perhaps more important is that 3-D printing is customizable. No one has to change the die casts to print a new cookie or candy. Just upload a new data file, and cookie number one is entirely different from cookie number two.

It's the idea of customizing that has reignited interest among the food companies. If a 3-D printer is flexible enough to make different cookies at every printing, it's also flexible enough to make me a cookie that is different from yours. Once we move beyond cookies to, say, breakfast bars, we're talking about a quasi-staple rather than an occasional indulgence. That opens the door to personalized nutrition. Your bar might have more protein and more vitamin A than mine. My bar might have more calcium and omega-3 than yours. While we're at it, each of us can print our bars in the shape of our alma mater's logo.

Lipson said that 3-D printing can, for perhaps the first time, combine cooking and information technology. After all, this is an age when our watches are continuously tracking our heart beat, blood pressure, and sleeping rhythms, and when we can order individualized DNA tests off the Internet. These data could be used to customize breakfast granola or even make pharmaceutically enhanced candy bars that contain the right dose of allergy or cholesterol medication.

Lipson's machines aren't limited to printing just candy bars and other sweets. In fact, some commercial models, like the Foodini, have brought flexibility to market. Essentially, any ingredient that can be made into a liquid, gel, or paste can be printed. That includes bread batters, cheese, pasta, jam, and carrot puree. Custom-shaped ravioli with garlic-infused ricotta? No problem.

In fact, Lipson sees 3-D food printing evolving much as regular paper printing did. Many color paper printers use only three base colors: red, green, and blue (RGB). When used in different amounts and combinations, these three colors can produce most of the colors we see. Similarly, 3-D food printers eventually might use only a handful of base flavors that, when mixed in various combinations, create a myriad of familiar, and even unfamiliar, flavors and textures. Lipson and his team have made purple cubes that taste like broccoli, which wasn't terribly popular among the taste testers.

Three-dimensional printers can be used to make food more conveniently. They can also be used to make food healthier, particularly when we consider the possibilities of personalized nutrition. Printed food can also be tasty. Convenient, healthy, *and* tasty. Those are three words that don't often go together. Potato chips are tasty but not particularly healthy. Kale is healthy but, at least in my opinion, not very tasty. We typically must choose one or two of these characteristics when selecting food, but finding foods that give us all three is difficult. Technological change allows us to break through nature's barriers. It is at least possible to imagine 3-D printed food that is affordable, convenient, healthy, *and* scrumptious.

Lipson said people sometimes are skeptical about creating a machine that makes "processed food." But it isn't clear that's the right term. Printed foods can be made and eaten *fresh*. That means there's no need for the preservatives or stabilizers that typically are included in processed foods to increase their shelf life. Moreover, as Lipson noted, you can use whatever ingredients you want, for example, locally grown organic wheat flour or heirloom pea puree. You're probably never going to be able to print a salad, but that doesn't mean many other fresh foods couldn't be printed. Grandma's apple pie required a lot of mixing and manipulation, but few would call it processed.

The printers could be programmed to time the printing so that haggard parents have to spend only a little time laboring in the kitchen. So long frozen pizza and frozen ravioli. Load in a few cartridges or capsules of ingredients and you're a mouse click away from a fresh, delicious meal. While it might be tempting to lump 3-D food printers in with microwaves as the demonic devices that promote processed foods, they are anything but. Lipson's team has worked with chefs from the French Culinary Institute to create veggie-stuffed turkey and patterned cookie dough. The devices might allow the home cook to make artisanal delights that previously were the purview of seasoned chefs.

While some creations—like purple broccoli-flavored cubes— are strange and are unlikely to gain widespread appeal, aren't new victual finds a foodie fetish? Bacon- and wasabi-flavored ice cream? Seems like the sort of food porn I regularly see posted on social media. If yours is a unique creation that no one else can have, it becomes all the more alluring and high status. Indeed, what puts 3-D printed foods in a completely different category than standardized, processed foods are the uniqueness of 3-D printed foods and the ability to customize them.

Lipson believes that the personalization of food will be one of the catalysts that popularize 3-D food printing. He mentioned that his teenage son loves to experiment with the 3-D food printer. Lipson imagines a world in which his son can not only upload pictures of his dinner, as so many foodie types are wont to do, but in which he can also share his data file on social media so that friends can re-create, and even tweak, his 3-D printed food creations. If Lipson's son one day can upload his recipes, and in so doing give us the ability to perfectly replicate his dinner, then the same could be done by a Michelin-rated chef, our favorite celebrity, or a mommy blogger.

Food printing is still in its infancy. Only a handful of research groups around the world are working on the technology. The first-ever conference on 3-D printed food was just held in the Netherlands in April 2015. Much remains to be learned.

Much discussion in the 3-D printing community is about what it will take for most households to adopt the devices, whether for printing food or something else. The vast majority of efforts in 3-D printing today are focused on creating smaller, less expensive models that can print in plastic. Lipson, however, thinks many households will never have a regular use for such a creation. Rather (and he acknowledges he is in the minority here), he thinks 3-D food printing could be the new killer app. The ability to print fresh, customizable food is what might one day bring a 3-D printer into our homes for the first time.

Lipson likened it to video games. When personal computers hit the market, a few households quickly adopted them, but most couldn't see the need or applicability to their daily lives. But the "frivolous" gaming software that ran on the computers first brought them into our homes. We wanted to play Pong and solitaire, which was reason enough for a personal computer once the price was right.

I saw the same trend play out in a slightly different way in my own home. My mom decided to go back to college for a master's degree in mathematics when I was in junior high. At about that time we got our first computer (and printer—the kind with the continuous feed paper with the perforated, dotted holes). My mother had to write a computer program for one of her classes, and what did she choose to do? She programmed a hangman word-guessing game with special animation and music. Only later did I figure out how to use the device to more easily write essays and keep track of the monthly budget; eventually I used it to carry out statistical modeling in my own research. Food may be to 3-D printing what video games were to personal computing.

Just as kids were more likely to be interested in computer games than were their parents, they may also become the early adopters of 3-D food printing. As part of their outreach and education efforts, Lipson and his colleagues travel to elementary schools to show off their robotic inventions. Although the children like the robots, Lipson said they get really excited about the 3-D food printer. Aside from whatever tasty treats a 3-D printer might produce for us to eat, perhaps one of its most promising applications is fostering interest among children in learning about nutrition and food science.

There's still much to be done before most of us will want a 3-D printer in our kitchens. For one, they're slow. You'd have to start printing around lunch if you wanted dinner. One European food designer, who printed gold-embossed chocolate globes that encased more than six chambers with delicacies from around the world, remarked, "To prepare one per guest in a restaurant with 40 patrons would take almost two days of almost continuous printing. It's not very realistic. At the moment it's a way to show craftsmanship."[4] Moreover, a printed cookie or cake still has to be cooked in a conventional oven. Lipson thinks he has a fix for that—a laser that cooks the ingredients in real time as they're being printed. This sort of cooking technique, combined with different food textures and programmed pockets of air, have the potential to create tastes and textures unknown to even the van-guards of molecular gastronomy. A final drawback is that novel 3-D printing designs are today mainly limited to those people who can manipulate CAD (computer-aided design) software, which is tailored more for engineers than home cooks. More user-friendly interfaces will increase the appeal of 3-D food printing, and they're already starting to appear. For example, some home devices can scan an existing 3-D object and then reproduce it in plastic. It is a small step to instead scan an object and reproduce it in sugar or chocolate.

Perhaps someday soon we'll have a 3-D food scanner/printer sitting next to the coffee maker, food processor, mixer, and all the other time-saving devices that grace our kitchen counters.

That is, if we even need a home kitchen in the future.

My wife and I are big fans of Mexican and southwestern food. On a recent trip to Las Vegas, we jumped at the chance to eat at Mesa Grill, one of the trendy restaurants owned by the ce-lebrity chef Bobby Flay. Over a couple house-made margaritas, we thoroughly enjoyed an appetizer of goat cheese queso fundido—a wonderful chile-infused cheese dip served with amazing blue corn tortillas. We liked the appetizer so much that my wife bought one of Flay's cookbooks, and a few months later we tried to make the dish ourselves.

The recipe indicated it would take about a half an hour to prepare and cook the dish. By the time we located the right chile peppers, roasted and peeled them to make the vinaigrette, made the roux that served as the base for the Monterey jack and goat cheeses, and then baked the whole thing, we'd spent the better part of an afternoon. The dish turned out well, and our guests licked the platter clean. But I thought it wasn't quite the same as Mesa Grill's. Maybe it was the absence of the buzz flowing in from Caesar's Palace, where Mesa Grill is located, or maybe we didn't use the right chiles, or maybe we let the butter get a bit too brown. We'll never know. Guess we'll just have to wait till our next trip to Vegas to have the real deal.

Or will we?

Have you ever tried to replicate a dish from your favorite restaurant or chef? Even if you watched the chef make it yourself on the Food Network, and even if you have her cookbook with a detailed ingredient list and step-by-step instructions, who among us hasn't turned out a flop? Printing 3-D food works by taking a design—a computer recipe—and reproducing it in food. If you use the same ingredients and the same computer program, you'll get the same food. No flops.

Yet, as I've shown, 3-D food printing may be limited in what it ultimately can make. That's not to say it might not be useful to have one sitting on the kitchen countertop, only that it is unlikely to completely replace the kitchen. Replacing the entire kitchen will require an entirely different sort of innovation. How about a personalized chef that makes perfect dishes every time, without complaining, and without pay? Someone—or something—that can copy not just the food but the task of cooking itself? A robo-chef. If I can't get myself to Las Vegas, perhaps I can bring a robot version of Bobby Flay to me.

Mark Oleynik and his company, Moley Robotics, unveiled a prototype robot cook at the world's largest industrial fair in the spring of 2015.[5] The robot turned out perfectly prepared crab bisque. Perfect because the robotic arms were programmed to follow—in every way—the movements of a celebrity chef, Tim Anderson, who had won the British version of the reality television cooking competition *Master Chef*. Anderson had previously prepared the dish with a series of motion sensors attached to his hands and arms. The robotic arms were programmed to follow the chef's every move. The whole process is copied exactly—every intricate turn of the whisk, swirl of the spoon, and shake of the pan. The robot doesn't forget whether it added an ingredient, how long the onions have been sautéing, or when to reduce the heat. If both used the same recipe and ingredients, it would be impossible to tell the chef's preparation from the robot's. Actually, the robot would make the dish precisely the same every time, whereas even the best chef is known to make the occasional mistake.

In 1965 American women spent an average of 113 minutes per day (almost two hours) in meal preparation. By 2007 they were spending only 66 minutes per day, a 47 percent reduction.[6] The time spent in meal cleanup declined even more dramatically.[7] While we might have a tendency to romanticize the past, the reality is that cooking was often a monotonous, arduous, thankless task required to keep the family fed. Modern cooking and cleaning technology has given us more free time.

Of course, these technological changes have also brought some adverse consequences. More convenient, microwave-ready, easy-to-cook food in disposable containers often means more carbohydrates and fat than the homemade version. Yet the robot cook again opens the possibility that technological innovation may free us from nature's barriers. A robotic chef can cook virtually anything we can—healthy, fresh, or not. Letting the robot do the work, however, is much more convenient. That's particularly true if the robot can start work while we're commuting home from the office.

A robotic chef also can tackle a wide diversity of foods and cooking styles that we would never have time to master. Oleynik argues that we humans want unlimited variety and that we desire food variety—which is easy enough to confirm by gazing at the dizzying number of brands and flavors gracing the aisles upon aisles of supermarket shelves. But while we like variety in our diets—Italian today and Thai tomorrow—we may not know how to actually prepare all the things we'd like to eat. A robot's knowledge is limited only by the size of its hard drive.

Oleynik's robotic chef is more like a fancy kitchen countertop with arms than the rolling, talking robotic maid from the *Jetsons*. The prototype, which Oleynik said his team built in only four months, is all that's available at this point. Oleynik doesn't expect the commercial models to come out until 2017. The projected cost for the entire kitchen (which includes an oven, cooktop, sink, dishwasher, food processor, and two robotic arms) comes in at a hefty price tag of $15,000. That kind of cost will likely put a robocook outside the reach of most households. But that's only a bit more than what it costs to buy a high-end, commercial-style oven and cooktop made by companies like Wolf or Viking. Moreover, as Oleynik argues, price is ultimately a function of volume. The more units they make, the greater the savings. Keep in mind that many of our modern conveniences were much more expensive when they were first introduced. One of the first microwaves sold in 1955 for $1,300, which is more than $11,500 in 2015 dollars.[8]

The prototype robotic kitchen unveiled at the recent trade show consists of a long cylinder that looks to be about seven or eight feet long and sits atop a roughly three-foot-tall base. One side of the cylinder is open for viewing. The robotic arms are affixed to the top and backside of the cylinder; they move from side to side and backward and forward. On the countertop below the arms are pans and burners. To the left is a microwave and to the right a sink. The whole device would probably take up the amount of space occupied by a wall of cabinets in a typical kitchen.

The prototype works by a human placing ingredients, like butter and garlic, in containers in preassigned locations, where the robot arms know where to find them. Ultimately, the unit may contain a variety of internal cabinets with refrigerated or stored ingredients that it can retrieve as needed for the recipe. Right now the robot arms don't chop or cut, but Oleynik told me that giving the robotic hands a knife is not a big deal. And its eyes won't water after slicing into an onion.

Oleynik imagines a day when the kitchen will house a digital library of thousands of recipes by different chefs and restaurants. Moley Robotic's online promotional materials show beautiful, space age–looking displays that may one day accompany the machine. A simple touch of the button allows the user to sort by type of food, cooking technique, number of calories, user ratings, or any number of other criteria. It may ultimately be possible to record yourself preparing a dish, share your file on social media, and then hand the job over to the robot the next time. Or the sensors could record Grandma as she makes her famous apple pie. Many of us crave the dishes we had as a child, and now it would be possible—quite literally—to have that spaghetti marinara just like Mom used to make.

I asked Oleynik why I should have a robot cook in my home. He said it would be like having an experienced chef at my beck and call. The benefits are obvious if you don't or can't cook. But even if you're like me and enjoy spending time in the kitchen, sometimes we'd prefer to be in the adjacent room with dinner guests, or, after an exhausting day of work or travel, it would be nice to hand the chore off to someone (or something) else. And even those of us who like to cook rarely take time to prepare breakfast. The robocook can be set to start cooking before you're even out of bed. And even the best cooks are limited by what they've learned to prepare. A robocook is a multicultural, culinary polymath.

Only a month after his prototype was unveiled, I asked Oleynik if he'd received any negative feedback. Surprisingly, he said no. If the idea catches on and the technology advances, I suspect there will be some pushback. So I played devil's advocate and said, "Well, it doesn't seem natural. Shouldn't we be cooking ourselves?" Oleynik didn't buy it. He said if the robot works correctly, determining whether the dish was made by human hands or machine hands would be impossible. The outcomes are identical. That is, except that the human hands would be freed up to read, to hike, or play with the kids while the robot hands do the work.

Frankly, I find it a bit hard to imagine a day when most kitchens have a robot cook. But Thomas Jefferson probably couldn't imagine a day when his farm's slave labor was universally seen as not only morally repugnant but technologically obsolete. Abraham Lincoln probably couldn't imagine a day when almost every home had a device that kept food cooler than 40 degrees. Heck, in my own lifetime (I'm just a tad over forty), I couldn't have imagined that the percentage of households that owned a microwave would go from less than 10 percent to more than 95 percent today.[9] Our imaginations may not be a reliable guide for what technological progress can sometimes deliver.

Here's to imagining a day when a robot can not only cook but will do the dishes.

4

Synthetic Biology

BREWING MORE THAN BEER

There was a time—not too long ago—that reaching for a cold brew meant selecting from one of only a handful of behemoth beer makers. But if the choice for our granddads was between Miller and Bud, today we have those and more.[1] Many, many more. In fact, if we aren't satisfied by the overwhelming number of craft brews lining the shelves of liquor stores, it's easier now than ever just to make it at home. I've had some great (and some not-so-great) homebrews made by friends and students, and I've also seen what an overactive, explosive batch will do to the walls and ceiling of the laundry room. In the trendy fun of getting reacquainted with the sudsy delight, we often forget that beer is almost as old as the human race. In fact, it's been argued that without beer, we might not have developed civilization.[2]

Our ancestors, even those who created the thirteenth-century Bavarian beer purity laws, were employing, whether they knew it or not, a crude form of biotechnology. One view of biotechnology is that it is simply, as Merriam-Webster's online dictionary says, "the use of living cells, bacteria, etc., to make useful products." Brewers use a living organism—yeast—to convert one useful product (the sugars in grain) into another, more valuable product (alcohol). Even before brewers knew about yeast, they learned that sediment (and thus the yeast) from good batches should be mixed with the new. In this way they were selecting and manipulating microorganisms.

If yeast can convert sugar to alcohol, what else can it do? As it turns out, yeast is more than just an alcohol factory. Yeasts can eat up sugars to make flavors, fats, and fuels. And more. Yeast can make whatever its instructions tell it to make. By instructions, I mean the yeast's genetic code, or DNA.

My kids are Lego fans. But, especially when they were younger, I could immediately feel the anxiety build when they received for Christmas a box of Legos containing what seemed like a million pieces. The kids would want to play with the spaceship or car shown on the front of the box. But the only way for them to have it was for me to painstakingly go through the tiny instruction booklet page by page, piecing together each and every one of the tiny blocks.

Legos are standardized, but they come in a variety of shapes and sizes. If all the pieces were simply dumped in a pile in the middle of the floor (a common occurrence in our house, one with devastating effects on unsuspecting bare feet) with no instruction booklet, I could probably make an odd flying contraption, but the creation is certain to look nothing like the Millennium Falcon. That tiny instruction booklet makes all the difference. It tells me exactly how to combine all those disparate tiny pieces into a child's delight.

DNA is like a Lego instruction book. But rather than using pictures and a twenty-six-letter letter alphabet, DNA has instructions written in a four-letter chemical alphabet with the acronyms A, C, G, and T. Various combinations of these letters are the instructions that tell other molecules what to do and what compounds to make.

Yeasts' sets of instructions have evolved over time and in different ways, depending on their abilities to compete in different environments. But at some point some yeast evolved a set of instructions that tells its molecules to convert sugar to alcohol. Changing those instructions to assign it other tasks is the subject of the emerging field of synthetic biology.

At this point synthetic biology has no undisputed, consensus definition, but it involves the application of engineering methods and techniques to biology, particularly genetics. In some ways synthetic biology is synonymous with biotechnology, but synthetic biology is usually discussed in much more expansive terms. Inserting a gene from one species into another or manipulating an existing gene of a plant or animal is often couched in terms of genetically modified organisms (GMOs) and biotechnology. Synthetic biology is that and more. Scientists seek to create or manufacture entire strains of DNA that may not exist in nature; custom design genetic structures to program desirable outcomes in living cells; or use DNA as a storage device (one research team encoded all of Shakespeare's sonnets on a microscopic strand of DNA).[3]

If all that sounds a little scary, perhaps you will be somewhat comforted to know that many current applications of synthetic biology entail alterations in the lab to tiny organisms like yeast or bacteria, particularly *Escherichia coli* (*E. coli*). These little guys are used because they're so darn prolific. They replicate fast, and when they do, they create and carry with them perfect copies (or clones) of the new types of DNA the researcher desires. We're accustomed to eating foods made with yeast—whether it's beer, bread, or wine. But *E. coli?* Isn't that the stuff that kills you? There are types of *E. coli,* like O157:H7, that are indeed deadly. But most *E. coli* are not. In fact, we all have friendly *E. coli* in our guts, and we should be grateful for it because the bacteria create vitamins and help us digest food. There is little concern that the types of *E. coli* used in the lab for synthetic biology will cause illness because they don't even have the genes that would produce the toxins that harm us.[4] To confuse the two types of *E. coli* would be akin to saying Hitler and Gandhi were the same because they were both *Homo sapiens* and shared some of the same DNA.

In many biotechnology applications the researchers are interested only in using *E. coli* to replicate new DNA segments (the new DNA segments are pieced in and are replicated with the rest of the genetic material when the cell divides). After replication the new DNA of interest is cut out, and the remaining *E. coli* are discarded. So, even if you were concerned about using bacteria like *E. coli* for genetic research, you should know that these bacteria are not present when the researchers splice the new segment of DNA into whichever plant they wish to exhibit the new trait of interest.

But many researchers want to use the *E. coli* or the yeast for purposes beyond the mere replication, or cloning, of DNA segments. If the bacteria or yeast is given a gene responsible for producing a compound (say, a particular hormone or enzyme or fat), more bacteria mean more of the desired compound. And if the bacteria and yeast replicate quickly—as they are prone to do—then researchers can end up with a lot of the compound.

What if that compound is a substance used and badly needed by an estimated 4.9 million Americans?[5] What if that compound is insulin? Virtually all the insulin now taken by diabetics is produced by genetically altered bacteria in a lab.[6] Before that it had to be extracted from the pancreases of cows and pigs—a process that was both expensive and lacking in quality control.

As you might expect, the big pharmaceutical and biotech companies are actively involved in creating new bacteria and yeast that will produce desirable compounds. Eli Lilly was responsible for the commercialization of bioengineered insulin. Moreover, about 80 percent of today's cheeses are made with rennet (an enzyme used to help turn milk into curds and whey) produced by genetically engineered bacteria developed by companies like Pfizer and Chr. Hansen.[7] Before the development of the new bacteria in the 1990s, cheese makers often relied on rennet taken from the intestinal lining of baby calves. Researchers were able to take the bovine gene responsible for producing rennet and give it to the bacteria that now do the heavy lifting.

But a lot of the action in food-related synthetic biology is in companies you've probably never heard of. Silicon Valley start-ups like Muufri are trying to remake milk—without the cow. They've engineered yeast to produce casein, the key protein in milk, and are working on creating the other components of dairy as well. Solazyme is a San Francisco–based company (with manu-facturing facilities in Illinois) that is using algae to make oils that can replace dairy and egg fats. Other companies, like Chromadex out of Irvine, California, are working on bacteria that produce the antioxidants in blueberries and raspberries. Allylix is a San Diego company that uses genetically engineered yeast to make grapefruit and orange aromas and flavors. One of the leaders in the field is the Swiss company Evolva, which is already selling vanillin and stevia (a zero-calorie sugar substitute) produced by genetically engineered yeast. Evolva has its sights set on yeast that can produce saffron, which is the world's most expensive spice.[8]

Undoubtedly, many food consumers will regard some of these products as having a yuck factor. The intuitive emotional response, however, may fade once you learn a bit more about the motivations of the entrepreneurs and about the products the biological synthetics are replacing.[9] For example, much of the artificial vanillin on the market today is made with petroleum by-products. Is vanillin from yeast really more repugnant than vanillin from petrochemicals? The ingredients made by bacteria, yeast, or algae are often molecularly identical to those produced by Mother Nature. And why shouldn't they be? After all, the bacteria and yeast are using exactly the same genes to produce the vanillin, casein, or antioxidants as would the vanilla bean, cow, or blueberry. In fact, a friend of mine who works in the spice industry tells me that one way they know if a spice is synthetically produced is if its chemical profile is too perfect. Natural spices have flavor variations and imperfections. Natural spices also tend to be produced in areas of the world that do not regulate pesticide use as strenuously as does the United States. Lab-grown flavors don't have this problem.

The allure of the natural spices, fats, and proteins is strong. That is, until you understand the large amounts of resources—land, labor, fertilizer, and pesticides—required to produce these substances in the traditional way. The small size of yeast and bacteria shouldn't fool us into thinking that using these microbes might not have large effects. Some applications of synthetic biology are meant to get bacteria or algae or yeast, which do not feel pain or suffering, to produce proteins and fats that we typically take from animals, who suffer when they are improperly housed and cared for. For many vegans and vegetarians the ethical balance clearly tilts in favor of using bacteria and yeast rather than cows and pigs. Synthetic biology is also likely to appeal to environmentalists concerned about water, pesticide, and land use.

However, I must note that just as beer-making yeast require barley to produce alcohol, other lab-grown yeast and bacteria need feedstock and calories (corn is a likely source) to make insulin, vanillin, or antioxidants. Thus the net effect on resources and the environment may be less pronounced than we might have hoped. Moreover, just as beer production leaves behind so-called distillers grains, which are fed to livestock, many synthetic biology applications will also leave behind spent grain and waste products that might serve as a relatively inexpensive source of animal feed. Whether these new ways of producing flavors and foodstuffs are more or less resource intensive will depend on how efficient the bacteria and yeast become. Market prices will help us identify which is most resource intensive. To the extent that biotechnology applications can pass the market test, they may also address the concerns that many social justice advocates express about exploitative plantations, high food prices, and unequal access to quality, nutritious foodstuffs.

Some of the most exciting developments in food bioengineering aren't even among the Silicon Valley–like start-ups. They're being conceived by kids who haven't even finished high school or college. For more than a decade students around the globe have been assembling for an annual competition once hosted by MIT but now put on by the nonprofit International Genetically Engineered Machine (iGEM) Foundation. iGEM has become the premier competition in synthetic biology for graduate, undergraduate, and high school students.[10]

Twenty-five hundred participants on 243 teams entered the 2014 competition, which was held in Boston. High school teams from Texas to Taipei came, and college teams from Germany, Indonesia, China, France, and Brazil competed. Not all the teams worked on food or agricultural applications, but many did. For example, in an attempt to reduce food waste, Colorado State University students worked on bacteria that could break down used frying oil to create value-added products. Worried about pesticide use and losses in farmer income, a team from Wageningen University in the Netherlands engineered soil bacteria to sense and prevent the growth of a type of fungus known to damage banana trees. One French team sought to make *E. coli* smell like lemon, and another French team from Paris worked on a bacterium to fight body odor (a fix I would have gladly used on more than a few Parisian metro riders). A team from the University of Texas in Austin worked to give coffee drinkers a bacterium that will reveal the amount of caffeine in their morning latte. The team from the University of Maryland engineered a bacterial sensor for an oyster disease common in the Chesapeake Bay. Worried about malnutrition, Purdue University students presented a biotechnique to increase the iron content of staple crops.

One of the most intriguing entries came from a team from the City University of Hong Kong. The problem these students chose to tackle is one of the thorniest, most pressing problems of our time: obesity. More than a third of American adults and 17 percent of children are today obese. Almost 70 percent of us are overweight.[11] This global phenomenon is associated with problems like heart disease, stroke, and type-2 diabetes. With the annual costs of obesity for Americans estimated as high as $209 billion, solutions are badly needed.[12]

Easy solutions have been elusive. In fact, a lot of what we believe about obesity, according to an article in the *New England Journal of Medicine,* is myth or unsubstantiated presumption.[13] For example, small reductions in caloric intake over time are unlikely to lead to large reductions in weight, and eating (or skipping) breakfast has no clear effect on our waistlines. Obesity is a complex phenomenon, and there is no consensus on the causes or the efficacy of the proposed solutions. The increase in obesity has been blamed on everything (with varying degrees of credibility) from farm policies, antismoking campaigns, the popularization of food processing, working mothers, food advertising, better roads and automobiles, fast food, air conditioning, and food stamps to high fructose corn syrup, just to name a few suspects. The research suggests that most of the proposed policy solutions, like soda taxes and vegetable subsidies, are unlikely to have a substantive effect on weight.[14] That shouldn't be too surprising. After all, Americans voluntarily spend more than $60 billion annually on weight-loss products and services, and judging by the national obesity statistics, most of that doesn't work either.[15]

We need some fresh ideas. I tracked down two undergraduate members of the gold medal–winning Hong Kong iGEM team, Paul Tse and Marco So.[16] Even in a place like Hong Kong, obesity is a pressing concern, with 47 percent of men and 28 percent of women classified as overweight or obese.[17] Paul, Marco, and their team members approached the problem by recognizing that a variety of factors have led to increased caloric consumption, which in turn contributes to obesity. When people consume more calories than they expend, their bodies store fat. The most calorie-dense macronutrient we consume is fat, which provides more than twice the number of calories per gram as protein or carbs. They looked to *E. coli,* which already lives in our gut, and wondered whether it might be modified to eat up some of the fat from food that would otherwise end up as love handles. Paul said their first goal was to create a version of *E. coli* that would "absorb as much fat as possible." Second, they needed the *E. coli* to turn all that fat into something that might improve people's health.

The team decided to try to coax the bacteria into manufacturing alpha linoleic acid, an essential omega-3 fatty acid found naturally in flaxseed, walnuts, and soy. It is an essential fatty acid because our bodies need it yet cannot manufacture it. Thus it must come from our diet. Our bodies can convert the fatty acid into docosahexaenoic acid (DHA), also found in fish oil. According to Paul, DHA "has many benefits to humans such as improvement of vision and memory, and reduces the chance of suffering from cardiovascular diseases." The next time you're in a grocery store, take a peek down the aisle selling baby formula. You'll find the brands advertising the inclusion of DHA selling for hefty premiums. The purported benefits of DHA are so numerous that Marco called it "liquid gold."

This isn't the first time researchers have thought of promoting a type of *E. coli* for human health—as a friendly probiotic. As early as 1917 a German scientist isolated a type of *E. coli* that improves gastrointestinal health.[18] This bacterium, and others like it, are used today to help with irritable bowel syndrome, constipation, and other gastrointestinal problems.[19]

The Hong Kong team sought to create an entirely new bacterium. Marco described the process to me. He said they added some new plasmids to *E. coli*. A plasmid is a short circular DNA molecule inside a bacterial cell that is separate from the chromosomal DNA. Marco said you could think of it as akin to "inserting several CDs into the computer so the computer can do several new things." Marco said the particular plasmids they used were from deep sea microalgae.

Both Paul and Marco recognize that the general public is skeptical of the kind of work they are doing, but both are optimistic about the future. The team delivered talks to the general public and to high school students to introduce them to their project, and they constructed a website to detail their work and achievements. They found, particularly with their friends and family, a lot of support when they could explain what they were doing and why. "By explaining to my family and friends what synthetic biology is and how genetic engineering is carried out," Paul said, they more clearly understood what he was doing and became "less resistant to genetic modification." He said that synthetic biology is a tool. It could be an "evil tool to create biological weapons" or it could fight obesity and "elevate our quality of living." The key, according to both Paul and Marco, is not to reject synthetic biology outright but to identify appropriate regulations, surveillance rules, and safeguards and recognize that it is important to engage with the public—to be transparent and to educate and inform.

As for the future, the team envisions further tests, first with animals and then in clinical trials, and perhaps ultimately a commercial application. They're writing business plans and seeking funding. Paul says they hope to package their product as a "probiotic for slimming." As he sees it, we routinely buy products—like yogurt—that contain bacteria that we hope will make us healthier.

The Hong Kong team may not have invented *the* solution to obesity, but they may have taken a small step in the right direction. After all, few of us gain thirty pounds overnight. Obesity is the cumulative result of many small choices influenced by myriad factors. As such, we probably shouldn't expect a grand one-time fix. Rather, we should appreciate the many small steps that bring us back down to size.

To be sure, eating a genetically engineered bacterium—even if it has only new plasmids and not a new genome—is likely to make a few folks queasy. Can synthetic biology solve some of our food problems even if we don't want to eat genetically engineered superbugs? You bet.

I was able to catch up with Sarah Ritz, a senior biochemistry major, and Aaron Cohen, a senior biomedical engineering major, both of whom were members of the grand prize–winning iGEM team from the University of California at Davis.[20] When their team assembled in April (before the competition the following fall), they produced a lot of ideas during their brainstorming sessions. Because synthetic biology is such a broad field, the iGEM competition places few constraints on the teams' projects (however, iGEM does host a DNA registry repository—an inventory of Lego-like biobricks—that has thousands of biological parts used by previous iGEM teams and labs that competitors can order and use). Ritz and Cohen and their teammates toyed with several ideas, but they wanted to play to the strengths of their university and its location. California is the largest agricultural state in the nation, and UC Davis is one of the premier agricultural universities.[21]

One of the team's advisers was the research director of the olive center at UC Davis, and she alerted the group to an important problem facing the industry, which has expanded U.S. olive oil production by tenfold since 2007. California is the only state in the nation to produce olives commercially.[22] In 2013 it produced more than $136 million worth of olives, but most olive oil purchased by Americans has been imported from Europe. According to Ritz, as much as 70 percent of the olive oil imported into the United States is rancid by the time it reaches the consumer. Rancid oil has gone stale. It isn't necessarily harmful or even bad tasting to the average consumer. In fact, the UC Davis team conducted some blind tastes with consumers and found that many people actually preferred the rancid oil to fresh oil—perhaps because it is what they have become so accustomed to eating. Ritz said that fresh olive oil creates a tingling feeling in the throat—a phenomenon unfamiliar to many American consumers. Being habituated to blander, stale oil has its costs. Rancid oil does not have the same healthy compounds—like antioxidants—that are associated with fresh olive oil.

Concern about the quality of olive oil became so great that California lawmakers got involved. Ritz, Cohen, and the team visited olive farms and attended legislative hearings, where concerns were expressed about the unregulated oil imports from Europe. One of the contentious issues is that oils from different locations have different chemical properties, making a single uniform standard difficult to enforce. Nonetheless, in the fall of 2014, as the Davis iGEM team was heading to competition, the State of California adopted new standards for olive oils.[23]

The challenge is that testing oil quality is a tricky and expensive business—so much so that it may be cost prohibitive for small producers and sellers. Moreover, rancid oil is only loosely defined by California as oil that has an excess accumulation of smelly compounds. As a result, rancidity is rarely checked, and when it is, trained human tasters do the testing. But as is the case with any human endeavor, subjectivity is involved. It was precisely this conundrum that drew the attention of Ritz, Cohen, and their teammates. Would it be possible to create a *quantitative* measure of rancidity that was both inexpensive and quick to administer? They thought so.

Once classes ended, the team devoted the summer to the project. In fact, Cohen had been waiting for years for a chance to be on the team. He had grown up in Los Angeles and didn't head to college right away. Intrigued by the possibility of using biology to solve some of the world's most pressing problems, Cohen eventually enrolled in community college and tried—without success—to start an iGEM team. At an age when many of his peers had college in their rearview mirror, Cohen finally made his way to UC Davis and learned, to his surprise, that he had to try out for a spot on the team. He passed muster and jointed Ritz and five other students in earnestly applying what they knew about biology to the problems facing olive growers.

One of the first things the team had to do was create a more objective and quantitative measure of rancidity that went beyond "smelly." They focused on compounds known as aldehydes, which are responsible for the bad smells that arise from rancid oil (and the pleasant smells that come from cinnamon, vanilla, and cilantro). Their approach was to build an electrochemical biosensor that would detect the presence of particular types of

aldehydes and then send a signal that humans could easily ob-
serve if too many "bad" aldehydes were present.[24]

They took inspiration from the glucose monitors that diabetics use. These monitors require the user to place a drop of blood
on a strip that contains enzymes that recognize glucose. When
the enzymes recognize glucose, they send a biological signal that
can be translated into a digital readout that reveals the amount of
blood sugar. The trick for the students was to create an enzyme
that detects aldehyde rather than glucose and to come up with a
way to translate the enzyme's signals into a language that humans
can easily understand.

Enzymes are proteins produced by DNA, so the team looked
for segments of DNA (or genes) that, when placed in *E. coli,* produce enzymes that are sensitive to the presence of certain aldehydes. After reviewing the literature and some trial and error, the
team identified DNA sequences that produce enzymes that are
suitably sensitive to aldehyde. The final step was converting the
enzymes' signals into a digital readout indicating the presence of
three types of aldehydes and, finally, a quantitative measure of
the degree of rancidity.

The final prize-winning device works much like a blood
sugar meter. Place a drop of olive oil on a test strip and wait
for a digital screen to provide a reading. Ritz and Cohen said
their device really works but needs some fine-tuning to increase
its accuracy before it's ready for production. Whether California
olive growers eventually use the students' invention depends on
whether state law ultimately defines rancid standards in a more
quantifiable manner and how the market for fresher-tasting olive

oil grows. In the meantime the students will hand the project back to their faculty advisers for further refinement.

As this book went to press, Sara Ritz was thinking about going to medical school and was interning at a hospital. Aaron Cohen was wrapping up his college career and looking for a job in biomedical engineering. Both were optimistic about the promise of synthetic biology but realistic about its limitations and the potential challenges stemming from consumers' concerns. Ritz said that it was important to create awareness of the problem they were trying to solve. In their blind taste tests with consumers, most consumers were surprised to hear that rancidity is a problem but readily recognized the need for a test after learning about the issue. Some consumers expressed concern about biotechnology or GMOs, but the team's particular application didn't worry them. After all, the biotechnology was only in the sensor—it never touches the oil that we eventually eat.

Cohen thinks that the world may slowly come around to the concept of genetic engineering. If it does, it may well be because of the benefits we start to see from the new devices and foods that are created by our children and grandchildren who want to solve some of the problems we helped create.

5

Growing Flintstones

L ike a lot of parents, my wife and I have a son who was a bit of a surprise. By the time my wife realized she was pregnant, the end of the first trimester was nearing. We dusted off the baby and child-rearing books we'd kept around after our firstborn. Even a casual perusal of the genre reveals a pattern common among baby-book authors: scare the wits out of prospective parents.[1] Having not expected the arrival of our newest family member, my wife, Christy, had not been taking the prenatal vitamins that all the books suggest. She was horrified when she read that if she failed to consume enough folic acid—a B vitamin—the baby could develop spine and brain defects. When she relayed this concern to her father, he helped put things in perspective with a couple of questions. Were the pioneer women settling Kansas in the nineteenth-century taking folic acid pills? Were the Native Americans before them?

I'm happy to report that our son, now ten years old, turned out just fine. Our fears about folic acid reflect a relatively recent concern, historically speaking, about vitamins. Around the end of the nineteenth-century, nutritional scientists were just beginning to realize that good health depends on more than just how many fats, carbs, and proteins someone eats. Scientists soon discovered a variety of compounds that they ultimately called vitamins. One of the early pioneers, Elmer McCollum, an agricultural scientist at the University of Wisconsin who worked with dairy cows, helped determine the importance of vitamins. He also did much to hype their benefits, calling them essential "for the preservation of Vitality and Health," ultimately leading to a health food craze historians have dubbed Vitamania.[2] Food manufacturers quickly capitalized on fears of inadequate vitamin intake by marketing all kinds of supposedly protective, vitamin-enriched foods. The fears spawned all sorts of beliefs, quack diets, and eating advice—some promoted by McCollum himself—that persist in various forms today.

The good news is that those of us fortunate enough to live in a country like the United States have little to fear from vitamin deficiency. Food companies regularly fortify processed food with vitamins and minerals to foster the perception of healthfulness. The federal government requires fortification when there are concerns about the effects of deficiencies; examples are adding iodine to salt, which helps people avoid thyroid problems, and adding vitamin D to milk, which helps head off osteoporosis. Inexpensive vitamin supplements take up aisles in virtually every grocery store. Heck, we can even buy gummy vitamins shaped like Fred Flintstone that our kids will beg for.

Alas, many millions of people around the world are not so fortunate. An article in the British medical journal *The Lancet* estimates that 3.1 million children younger than five die every year from malnutrition, and 165 million children suffer stunted growth. Almost half of all childhood deaths in the world result from inadequate nutrition.[3] Even when people have enough calories to consume, available diets in many parts of the world lack essential vitamins and minerals.

One of the most prevalent problems is anemia. According to the World Health Organization, 47 percent of children younger than five (293 million) and 41 percent of pregnant women (56 million) globally are estimated to suffer from anemia.[4] In Africa and Southeast Asia, an astounding 65 percent of children younger than five have anemia. The red blood cells of anemia sufferers malfunction, resulting in fatigue, developmental disorders, and even death, particularly among pregnant women and babies. While anemia has a number of interrelated causes, insufficient dietary iron intake is thought to be a primary factor. More than sixty-nine thousand people died from iron deficiency anemia in 2010.[5]

Inadequate intake of other minerals, like zinc, is estimated to cause another ninety-seven thousand deaths. Almost a third of school-age children between the ages of eleven and fifteen globally lack sufficient iodine in their diet, resulting in significant health problems like goiters and brain defects and even leading to death.[6] In addition to lacking minerals in their diet, many people around the world are unable to secure adequate amounts of key vitamins.

Vitamin A deficiency is a serious global problem. In 2010 almost 120,000 people died prematurely and 108 million life years were lost—because of inadequate vitamin A intake. Even among the living, vitamin A deficiency can have serious adverse consequences. For example, in Africa more than 2 percent of children younger than five and more than 9 percent of pregnant women are believed to suffer from night blindness as a result of vitamin A deficiency. In Southeast Asia, where staple foods like rice lack vitamin A, it is estimated that 3.8 million pregnant women have night blindness. More than 125 million Southeast Asian children younger than five are believed to suffer from vitamin A deficiency, causing between 250,000 and 500,000 cases of childhood blindness.[7]

As I mentioned earlier, many people in the world do not consume enough folic acid and other B vitamins to maintain good health.[8] Vitamin D is also underconsumed. A 2008 article in the *American Journal of Clinical Nutrition* argues that "vitamin D deficiency is now recognized as a pandemic," causing rickets in children and osteoporosis in adults.[9]

Clearly, "hidden hunger" caused by a lack of vitamins and minerals takes a heavy toll. Malnutrition causes illness and death and keeps people trapped in unproductive economies. Research shows that malnutrition impairs the growth and cognitive ability of children. It also affects the amount of time people are able to work and the quality of their work. As a result, countries that have high rates of malnutrition also tend to have low rates of productivity and economic growth.[10]

The good news is that the solution to the problem is, at least in principle, straightforward: increase the amount of micronutrients available in the developing world. It's an approach with immense possibility. A few years ago the Copenhagen Consensus, a group of economists and researchers (including several Nobel Prize recipients), calculated that providing supplemental vitamins to undernourished children would be the world's best investment: yielding more "bang for the buck" than climate change interventions, heart attack prevention, or improving rural water supplies.[11]

Providing vitamin supplements (think Flintstones Vitamins on a global scale) has indeed produced positive outcomes in many parts of the world. The approach, however, has proved less beneficial than the optimists had predicted. Vitamin supplements present a number of challenges. First, you've got to deliver them to where they're needed—some of the most remote, unpaved, undeveloped places in the world. Then you've got to convince people to take them. Regularly. Then you've got to do it all over again. Every year. In perpetuity. Supplements are a one-off, partial solution to an ongoing problem. Writing in the journal *Nature,* one group of scientists argued that "despite numerous efforts to tackle vitamin and mineral deficiencies through supplementation, industrial fortification or dietary diversification, deficiencies remain widespread among two billion people."[12]

A more innovative, bottom-up approach is starting to challenge this top-down approach to ending malnutrition. One of the root causes of malnutrition is lack of dietary diversity, caused by both a lack of access and the inability to afford different food-stuffs. Many of the poorest people on Earth get the vast majority of their calories from a single staple crop. The staple crop that prevails in a given region depends on climate and custom, among other factors. In some places it's rice, in others beans, in others corn (or maize, as it's called in many parts of the world), wheat, cassava, or sweet potatoes. Although these crops may be agronomically suited for the regions in which they're raised, insofar as delivering a punch of calories, no single crop can supply all the nutrients people need. In the developed world, we beat malnutrition by eating a diversity of foods: oranges provide vitamin C, meat provides iron and B vitamins, carrots and sweet potatoes provide vitamin A, processed breakfast cereals are loaded with multivitamins, and kale, well, what micronutrient doesn't it provide?

The world's poor not only tend to rely on a single staple crop for sustenance but often grow their own food. It has been estimated that 2.5 billion people worldwide live from agricultural production and 70 to 80 percent of farmers in sub-Saharan Africa and in Asia are smallholders who barely meet their subsistence needs.[13] The diets of many of these farmers are limited to what they can grow. They save some seed from the previous year's harvest, or trade with locals or quasi-government agencies, to plant the fields again next season.

In this conundrum may lie a solution. If the staple crops of these farm families were more nutrient dense, some of the problems of malnutrition could be solved. Biofortification is the science of breeding crops to increase nutritional content.

Some crops naturally produce trace amounts of vitamin A or B, and natural genetic variation within a given species means some varieties produce more potent nutrients than others. This natural genetic variation allows scientists to use conventional breeding techniques to selectively breed higher nutrient content into the varieties already being grown in the field. The process can be hurried along by using marker-assisted breeding, which involves using DNA tests to identify the genes responsible for expression of micronutrients and then selecting varieties with the right genes for conventional breeding. Some crops, however, do not create some micronutrients. In these cases biotechnology can help. By introducing new genes that instruct the cells of maize or rice to make vitamins A or B, it is possible to give subsistence farmers the power to grow vitamins themselves. Thus genetic engineering offers entirely new possibilities for biofortification and can dramatically speed up the process. The minimum number of crop seasons required for scientists to create biofortified crops using conventional breeding methods ranges from seven generations for clonally propagated crops like potatoes to seventeen generations for cross-fertilizing crops like corn; by contrast, with biotechnology, once a gene is identified and inserted in a variety, it is ready in a single generation.[14]

The power of biofortification—regardless of whether it comes from conventional plant breeding or through newer technologies—is that it puts the micronutrients in a place that they

replicate naturally, year after year, in a vehicle families are accustomed to eating. Once subsistence farmers gets their hands on
a crop variety with higher nutrient content, the fix perpetuates
itself as they save seed for the new crop year. The strategy behind
biofortification is a one-time delivery of seed that continues to
provide nutrients with each new harvest. For some crops (like
maize and pearl millet), farmers will want to buy new hybrid
seeds each year to maintain their yields. They will need to plant
new varieties and other crops (like sweet potatoes) each year to
protect against plant diseases. Even still, once biofortified crops
are incorporated into existing seed supply chains in these countries, their benefits have long-lived potential. Of course, we'd
really like these folks to have access to a diverse mix of nutrient-
dense crops, but we have to start somewhere and work up. Utopia
doesn't arrive in one package.

A science aside. Plants can create vitamins but not minerals. Nonetheless, different plants have different proclivities for
mineral availability. That is, some plant varieties (or genetically
engineered variants) can increase their mineral availability by
releasing iron or zinc bonds with other molecules in the plant,
making the minerals more or less accessible for human digestion.
Other varieties promote mineral uptake through roots. But, tech-
nically speaking, plants don't make minerals. They can manufacture vitamins from scratch.

*   *   *

Abdul Naico first walked through my door in Stillwater, Oklahoma, in 2007. He was an eager graduate student on a Ford

Foundation Fellowship from Mozambique. In my department of agricultural economics at Oklahoma State University, we typically have about sixty-five graduate students enrolled from all corners of the world. I've advised students from Haiti to Germany, from South Korea to Kansas, from North Dakota to Florida, and lots of places in between. With it comes the thrill of meeting new people, learning new customs, and appreciating the unique ways that make the world turn. The downside is that many new students are looking for research funding without a passion for or clue to what they want to work on.

Abdul Naico was different. From the minute he sat down in the chair across from mine, he knew the problem he wanted to tackle. He came from a country where malnutrition was rampant—44 percent of children in Mozambique have stunted growth and 33 percent of the childhood deaths are attributable to malnutrition.[15] For many of us, death is something rare and shocking, seldom experienced except when we have an elderly parent or grandparent. Murder makes the nightly news because it is so relatively rare in our culture that it captures our attention and our sympathy. Naico returned home during his graduate studies to help bury his brother, something he rarely mentioned and that I learned about only after other students told me what had happened. Life expectancy in Mozambique is about twenty-five years shorter than it is in the United States.[16]

The problem Naico focused on was vitamin A deficiency in Mozambique. Before coming to graduate school, he'd worked with plant breeders and government officials who had introduced several new varieties of orange-fleshed sweet potato suitable for

the climate in his country. Sweet potatoes are an important staple crop in Mozambique, as are maize and cassava.[17] Unlike the
sweet potatoes familiar to Americans, the typical variety grown
in Mozambique has white flesh. Although the white sweet potato
is hardy and calorie dense, it misses key vitamins, notably vitamin A, that also are lacking in the diets of many Mozambicans.

Back in 2000, the government of Mozambique undertook
the production and distribution of orange-fleshed sweet potatoes
to try to improve malnutrition and food security. Orange-fleshed
sweet potatoes provide an inexpensive source of beta-carotene,
which the body coverts to vitamin A. Despite a government-
funded awareness campaign promoting the nutritional benefits of
orange-fleshed sweet potatoes, the traditional white varieties remained the dominant choice for Mozambicans. Naico wondered
why that was the case and sought to figure out how to convince
farmers to adopt the new orange varieties.

During the summer of his first year in graduate school,
Naico returned home to figure out why the new orange sweet
potatoes had not taken hold. He was armed with new statistical
and survey techniques and set up shop at local markets in urban
and rural areas to interview buyers as they entered the market.
He gave people choices of different bags of sweet potatoes that
varied in the color of the flesh (white or orange), size, firmness
and moisture content, and price. By strategically varying the
types of choices presented to participants, he was able to sort out
which attributes were most important to consumers and what
they were willing to pay. He also wanted to better understand

what people knew about the health benefits of orange sweet potatoes, so some participants were randomly given information about the vitamin A content of the two sweet potato varieties and others were not.

One problem that arises in surveys of this sort is that people will lie (intentionally or not) about what they do when the researchers aren't watching. Over the years we've found that consumers will often *say* they're willing to pay two to three times what they really are willing to spend. The disconnect could be the result of social desirability bias (we want researchers to think highly of us so we present ourselves in a positive light) or even self-deception that goes unchecked by a real budget constraint (surely I'm a good and friendly guy who'd buy the excellent widget this researcher is selling). Naico knew the research on these topics from his coursework, so he had planned to give consumers a small fee for participating and then ask them to actually buy the type of potato they had chosen. The trouble came when he started handing out participation fees. Lines grew and violence eventually ensued when Abdul implemented his plan to use only every third respondent (to ensure a random and representative sample). I got a frantic email after one near fist fight. I learned a lesson not only about sweet potato preferences but also about how much easier consumer research is in the developed world. Naico settled on a compromise strategy. He wouldn't pay people to participate or make them actually pay for the potatoes, but to give people an incentive to think carefully, he'd randomly pick out one of their sweet potato choices and give it to them.

After weeks in the hot sun and more than three hundred interviews with consumers, Naico was able to glean a few lessons. Information mattered, particularly in rural areas. People given information about the nutritional content of orange sweet potatoes were more likely to pick orange over white. Despite the government's educational efforts, not many people understood the benefits of the new orange sweet potatoes. Another key insight was that taste (as reflected in preferences for firmness and moisture content) trumped everything else. Yes, consumers tended to prefer the orange varieties to the white but only if they were firm and tasty.

Abdul Naico found at least a partial answer to the question that had motivated his graduate studies. The new orange sweet potatoes weren't competing well with the white varieties because they lacked firmness (or ample dry matter content). He returned home with a new mission to help policy makers and plant breeders understand that the nutritional benefits of biofortified sweet potatoes will be realized only when the new varieties are just as large and tasty as the old. Today he is in Mozambique working for the International Potato Center, where he is seeking to improve his nation's health by expanding the production, distribution, and consumption of orange sweet potatoes. Careful studies of the impact and effectiveness of the orange sweet potato programs in Uganda and Mozambique have found that the delivery of orange sweet potatoes significantly increases vitamin A intake among children and women in a cost-effective manner.[18]

Naico's research showed that breeding a more nutrient-dense crop variety is not always a simple solution. Farmers must be willing to plant the new variety. It must be tasty and provide yields comparable to those of the old varieties. Thus research on biofortified crops requires both advances in genetics and crop science and a keener understanding of what the farmer-consumers want and how they decide what to plant.

This insight is taken seriously by HarvestPlus, an organization that started in 2004 with funding from the Bill and Melinda Gates Foundation and a global consortium of agricultural research centers that goes by the acronym CGIAR.[19] HarvestPlus has ongoing efforts in more than forty countries to develop and disseminate nutrient-rich seeds. The organization is involved in Naico's sweet potato efforts in Mozambique, vitamin A cassava and maize in Nigeria and Zambia, and high zinc wheat in India and Pakistan, just to name a few.[20]

The efforts at HarvestPlus have attracted international attention and talent. Take, for example, Gene Kahn. Thanks to Michael Pollan's influential book, *The Omnivore's Dilemma*, Gene Kahn is perhaps best known as the hippie-turned-organic-farmer-turned-corporate-executive. After creating Cascadian Farm and turning it into one of the largest organic brands in the country, he moved to General Mills as vice president and global sustainability officer. What is perhaps less well known is that Kahn quit to join forces with the Bill and Melinda Gates foundation and then HarvestPlus as head of Global Market Development. He's headed up initiatives to improve market access and distribution for nutrient-rich seed throughout the world. He's since directed the majority of his efforts elsewhere, but the work at HarvestPlus continues.

Given Naico's research, and other findings like it, Harvest-Plus invests not only in plant breeding but also in consumer research to identify which varieties will be most palatable in target countries. I've worked with HarvestPlus on a couple projects in Rwanda and Guatemala involving new beans that are high in iron.

Despite the high prevalence of anemia in Rwanda, the country has no iron supplementation programs for young children, nor are common foodstuffs in the country fortified with iron. But Rwandans eat a lot of beans. They have the highest per capita bean consumption in the world, and 97 percent of rural households in Rwanda grow beans.[21] New varieties of iron-fortified beans have been developed through conventional breeding methods, and research shows that consumption of the new beans can help with iron deficiency anemia. However, it is not yet clear whether Rwandans will want to plant and eat the new high-iron varieties.

The research team at HarvestPlus conducted research among more than five hundred Rwandan consumers, looking at preferences for a new white and red high-iron bean; a popular local variety had iron content that was 40 percent lower. The research showed that Rwandans liked the red high-iron bean more than the white high-iron bean. Both high-iron beans were preferred over the local variety. Moreover, consumers were willing to pay a premium for the new red high-iron bean. Information about the nutritional benefits of the new variety further boosted acceptance of the new variety, regardless of whether the message focused on the good (more iron improves health) or the bad (less iron deteriorates health). Additional messages at a later date resulted in consumers who were willing to pay even more for the new high-iron beans.[22]

HarvestPlus has used conventional breeding techniques to create high-iron beans, high-zinc wheat and rice, and vitamin A cassava and maize that hold the promise of improving and saving hundreds of thousands of lives by providing more nutritious foodstuffs. But conventional breeding has its limits. For example, rice is one of the most important staple crops in the world (along with maize and wheat). A report from the United Nations Food and Agriculture Organization indicates that almost half of the world's population relies on rice.[23] Yet no known productive strain of rice produces grain with vitamin A.

The German scientist Ingo Potrykus decided to take up that challenge more than two decades ago. Early in his career Potrykus helped develop a model for petunia experiments related to genetic engineering technology. When mounting evidence revealed crises of hunger and malnutrition in the developing world, Potrykus shifted his focus from the floral shop to the dinner table. In fact, he told me that his goal had always been to help address food security problems in developing countries, and for that reason he focused on cereal grains in the mid-1970s.[24] The experience of years of hunger in his youth was one factor that motivated his interest in applying the scientific knowledge he'd gained in plant science and genetic engineering to the problem of food insecurity and malnutrition.

"I worked with rice because this guaranteed the greatest impact," Potrykus said. "No other crop is as important for human nutrition, especially in developing countries." The idea of focusing on vitamin A came in the early 1990s when he realized that while hunger (a lack of calories) is a major problem, many more people in the world are hampered by hidden hunger (malnutrition and a lack of micronutrients). Potrykus said he never had any interest in "commercial exploitation of our invention. From 1991 onwards this was a humanitarian project." As a professor of plant sciences at the Swiss Federal Institute of Technology, he worked with a colleague, Peter Beyer, for almost ten years, trying to demonstrate that it is possible to activate the biochemical pathway for beta-carotene production in rice endosperm. They had many setbacks and frustrations along the way, but by 1999 Potrykus and colleagues had figured out a way to get rice to produce beta-carotene, which the body converts to vitamin A.[25] They did it by incorporating genes from a daffodil and a soil bacterium into rice DNA.

The rice they produced is called golden rice and gets its name from its color. Orange carrots and sweet potatoes get their color from carotene, and the presence of this compound ultimately provides the vitamin A. Because golden rice also has carotene, it also has an orangey, golden hue. The initial varieties of rice created by Potrykus and colleagues expressed only a small amount of vitamin A. Further iterations of golden rice have resulted in a twenty-threefold increase in the carotene content.[26] Current varieties can produce 55 to 77 percent of the recommended daily intake of vitamin A by eating a mere hundred grams of uncooked rice (or about half a cupful), and human research has found it safe and as effective as vitamin A supplements.[27]

The controversy surrounding genetic engineering—genetically modified organisms, or GMOs—has hindered and obscured Potrykus's work, but he blazed a path that could eventually lead to better nutrition around the world. A recent study in the journal *Nature Biotechnology* lists thirty-five different studies of nutrient-rich crops produced through biotechnology, including vitamin C–rich corn and potato; vitamin A–rich rice, potato, corn, wheat, cassava, and sorghum; folate (B9)–enhanced rice and corn; vitamin E–enriched corn and rice; high-iron rice, corn, and barley; and high-zinc rice and barley. Some researchers are working on multifortification approaches that combine many of these in one multivitamin crop.[28]

Not only are these crops technologically possible, but some, like golden rice, are ready for planting. And research shows consumers want them. Indeed, the vast majority of the consumer research studies on the topic have shown that people prefer, and are even willing to pay a premium for, nutrient-enhanced crops, even if they are biotech.[29] I've conducted two surveys of consumers in the United States, and even though Americans are less concerned about malnutrition, the majority of consumers say they are willing to pay a premium for golden rice over traditional white rice.[30] Another study in the Philippines found that if both varieties were identically priced, 60 percent of consumers would pick golden rice with the remaining 40 percent choosing conventional rice.[31]

These studies show that many consumers would like to have access to golden rice and other nutrient-rich crops created through biotechnology. However, fear, politics, and regulatory hurdles have kept these crops out of developing countries.

Potrykus officially retired from his academic post in 1999 when he was sixty-five. Today he's in his eighties. In his retirement he helped found the International Humanitarian Golden Rice Board, has collaborated with rice-breeding institutions throughout the world, and has even met the pope, all in an effort to combat vitamin A deficiency among some of the most impoverished people in the world. He is hopeful but realistic, telling me that he doubts he will have the pleasure of seeing golden rice planted in farmers' (and not just scientists') fields in his lifetime.

Potrykus has worked for decades to obtain regulatory approval for golden rice. Twenty years have been spent collecting data and materials to satisfy the regulators. The efforts have been hindered, in his assessment, by "extreme precautionary regulations" fueled by a "financially and politically influential ideology-based opposition." Reiterating his lack of interest in commercial rewards, he expressed frustration because every year of delay in deployment of golden rice results in millions of additional cases of blindness and death. The research supports his position. One recent study in the journal *Environment and Development Economics* estimated that the failure of regulators in India to approve golden rice from 2002 to 2012 led to more than $2 billion in economic losses and 1.4 million life years lost from inadequate vitamin A in that country alone.[32]

It is ironic that many of the initial first-generation biotech crops, such as herbicide resistant soybeans and insect-resistant corn, created by Monsanto, Dow, and Bayer, have been approved throughout the world, whereas these newer second-generation crops created by governments, academics, and nonprofits have stalled in regulatory limbo. The nutrient-rich biotech crops could arguably do much more good in the world than the original pesticide-resistant crops, but many of the entrepreneurs and inventors who have developed the biofortified crops lack the legal teams, political power, and financial resources to clear the regulatory hurdles.

Potrykus is hopeful that golden rice will be approved by regulators throughout Asia by 2019, two decades after its creation. I asked what worried him most with regard to the challenges associated with food, health, and the environment. As he sees it, "We have numerous, most severe problems ahead of us and we refuse to apply one of the most powerful technologies to help solving those problems." Reiterating the conclusions drawn by virtually all of the most prestigious scientific organizations in the world, he argues that "there is broad scientific consensus that the technology is at least as safe as all techniques we have been using. . . . There is not a single documented case of harm from the technology—I doubt that there is any technology ever with such an unprecedented safety track record."

Most would agree that malnutrition is a critical problem worth addressing, whatever the method used. The best solutions will probably depend on the particular circumstances and people involved. Sometimes providing inexpensive supplements may work well. Other times traditional crop breeding to beef up micronutrient expression in staple crops may do the trick. In other cases genetically engineered crops may be the best bet. It will take a combination of plant science, genetics, economics, politics, and even consumer research to experiment and find the best way forward. All tools should be on the table—including the possibility of new tools we haven't yet invented. Even better solutions to malnutrition may be waiting to be found by a humanitarian entrepreneur.

6

Farming Precisely

Although I spent my childhood summers working on neighbors' farms, I didn't learn much about how to run a farm. I was a hired hand. Move a tractor to the next field. Hoe the weeds here. Go to the co-op to buy Roundup. Those were my jobs. So I decided to plant a small garden in our backyard to teach my kids about where food comes from. The result was a calamity of errors.

I did not know the precise time of year each seed should be planted, so some crops never sprouted. Those that did seemed to get too much water at times and not enough at others. The insects arrived just about the time the produce was ripe. We were eventually able to produce a few scrawny carrots and potatoes. But before we had enough for a full meal, the watermelons took over. The whole plot eventually became a mess of watermelon vines. And weeds. And bugs.

When the watermelons looked like they might be ripe, we used a precise scientific method (thumping and listening) to judge when the melons were ready, except we picked several that were still pinkish-white on the inside. The rabbits or birds or deer seemed to know better how to judge ripeness, because we found holes exposing the bright red flesh. That is, when the melons weren't eaten entirely. I guess you can say that we learned a few things about how *not* to farm.

A seasoned gardener or farmer would no doubt manage much better than we did. But even they have a tough time getting things exactly right. In retrospect, one of the things we did wrong was to plant too many watermelon seeds too close to other crops. But what is the right seeding rate? And is there any reason to believe that the right rate is the same everywhere all the time?

Most food consumers have difficulty comprehending the enormous challenge of answering that sort of question for the typical grower who makes a living farming. For example, if you take all the cropland in the United States devoted to growing lettuce, half of that land would be on farms with more than 1,815 acres and half would be on smaller farms.[1] That means that the median lettuce field is managed by a farmer who has 1,373 football fields (including end zones) of lettuce to oversee. It's not just lettuce. More than half the land devoted to growing tomatoes is managed by farmers who have to care for 620 football fields of tomatoes. The figure is 688 football fields for wheat and 453 football fields for corn.

If I had a hard time keeping out the weeds and bugs and figuring out how much seed, water, and fertilizer to apply in a garden the size of 0.001 football fields, imagine the challenge for the lettuce farmer with more than a million times the amount of land under his stewardship.

Because it is physically impossible to be in all places at all times, the way most farmers have traditionally handled this challenge is to uniformly manage their fields. That is, they put out the same amount of seed per acre, at the same depth, with the same amount and type of fertilizer and insecticide for the whole field. It's a practice that David Waits, founder, former farmer and CEO, and current board chair of Site Specific Technology (SST) Software, called "tabletop farming."[2] A tabletop is a flat, uniform surface, and many farmers treat their fields like they are a flat, homogeneous plain akin to the top of an office desk.

The trouble is that farmland can be incredibly variable from one site to the next. As Waits put it, "A field is a highly variable surface." It is "a multitude of surfaces." A field isn't just a field. It is a combination of many smaller land areas that have different types of soil, soil nutrients and microbial populations, amounts of moisture and access to sunshine, and inherent proclivities for attracting weeds, insects, and diseases. As a consequence, when farmers tabletop farm, which is the way they always have, they apply too much seed, fertilizer, water, and herbicides in some parts of their fields and too little in others. That's true not just for the monoculture corn farmer but also for cover-crop–rotating rutabaga growers. Farmland variability is a fact of nature, not the choice of crop or a farmer's ideology.

It's a fact that's long been known. In ancient times (and still today in many developing countries), people farmed for subsistence and had few opportunities to make a living off the farm. In those settings famers could keep their fields small and their knowledge of every square foot intimate.[3] Given the need to affordably feed a growing urban population, and Americans' apparent lack of desire to give up their urban/suburban lives for rural living for more than a few vacation days a year, the United States is unlikely to see the problem of land variability solved by a mass movement back to near subsistence farming on a few acres.[4] The challenge lies in helping today's real-life farmers figure out how to prevent the under- and overapplication of farm inputs.

The overapplication of fertilizer, water, and pesticide has implications for the farmer's bottom line, but larger issues also are at stake. Environmental and public health concerns result from the flow of excess nitrogen, phosphorus, and herbicides from farm fields into underground drinking water and into streams and rivers. Excess nitrogen and phosphorous in rivers and lakes cause the growth of algae and plankton, leading to algal blooms that cover the surface of the water with green, brown, or red muck. Not only are these blooms ugly, they can sometimes produce dangerous toxins, and when the blooms become severe, they consume all the oxygen in the water, resulting in hypoxia. Hypoxia creates dead zones where fish and other water life cannot survive. The second-largest hypoxic zone in the world is in the Gulf of Mexico, and some experts believe that agricultural runoff is largely to blame.[5]

It's not as though a farmer wants to overapply fertilizer and pesticides. These are expensive inputs and their use comes at a high cost. For example, in 2014 a central Illinois corn farmer was likely to spend $173 per acre on fertilizer and $66 per acre on pesticides, even on highly productive farmland. All that was on top of the $119 per acre the farmer paid for seed. Of all the nonland costs associated with growing corn in Illinois, about 30 percent are tied up in fertilizer. All commercial crops use costly fertilizers and pesticides. Take an iceberg lettuce farmer in California. She'll spend about $550 per acre for fertilizer, $600 per acre on pesticides, and $175 per acre for seed. Even an *organic* leaf lettuce farmer in California will spend almost $800 per acre on fertilizer, $176 per acre on pesticides, and $148 per acre for seed.[6] If you've got a thousand of acres of farmland, it's easy to see how such costs can quickly escalate. Still, farmers often overapply fertilizer in part because it can act as a sort of insurance but also because, at least historically, they have had no technological means for being more precise about it.[7]

David Waits is helping solve that problem. Waits knows full well the challenges of being a farmer. He was one. After high school he gave college a try, but those were heady times with plenty of opportunities in rural America for those willing to take some risk. He thought, "How could I justify sitting in psychology class while people who needed my help were doing productive things back home?" Waits put down the books to be productive. He headed back to southwest Kansas to manage his dad's farm. Waits's father also owned a fertilizer business, and it gave Waits an intimate perspective of the input side of the farm business. Of course, in those days farmers had no mechanism for economically applying different rates of fertilizer throughout the field. They would apply the usual amount, or two times the normal amount, or none at all. Waits innovated, adopting center pivot irrigation and making other improvements, but when the oil boom began to bust in the 1980s, the rural economies of western Texas, Oklahoma, and Kansas took a big hit. After a decade on the farm Waits looked for other opportunities and returned to school.

With two kids in tow, Waits and his wife moved into student housing at Oklahoma State University in 1984. He didn't know it, but a "quiet revolution in agriculture based on information technology" was starting.[8] Massey Ferguson had just introduced a yield monitor—a device that measures in real time the number of bushels being harvested per acre as the harvester moves through the field—and was beginning to consider ways to make use of data on in-field yield variability. But linking these measures to geographic space was an arduous and time-consuming task. The global positioning system (GPS), for example, hadn't yet been invented. At about that time the concept of precision agriculture, as we today think of it, was brand new. The term itself was first used during an academic conference in 1990. As one article in the journal *Science* put it many years later, "Precision agriculture, or information-based management of agricultural production systems, emerged in the mid-1980s as a way to apply the right treatment in the right place at the right time."[9]

Although Waits was an economics major, he'd always been interested in maps and aerial photographs of farm fields, and he took some electives in geography. Toward the end of a course in cartography, a professor introduced Waits to the concept of geographic information systems (GIS). The professor explained how GIS could help city planners site a landfill to satisfy zoning restrictions that set minimum distances from houses, schools, and businesses. Waits had no interest in landfills. But he had been a farmer. He began thinking about how this technology could be applied to what he knew best.

Waits found a job with the Dairy Improvement Association that allowed him to continue his studies, eventually in graduate school with his attention focused fully on geography. He began concentrating on courses about remote sensing and other "geotechniques." Waits wanted to leverage his experience and knowledge in agriculture, but how GIS could be put to commercial use still was not clear. He said, "I knew I didn't know enough to capitalize" on the technology. That would require more study.

The challenges he wanted to address were interdisciplinary. The problems were not just geographic but required knowledge of engineering, agronomy, and economics, among other fields. He recognized the need for a degree program that would provide those skills, and he joined the interdisciplinary doctoral program in land use planning at Texas Tech University, where he was able to fashion his own program focused on rural land management. So while he was learning more geotechniques and studying the remote sensing techniques applicable to agricultural landscapes, he was also learning about environmental law, environmental impact analysis, range and wildlife management, agricultural economics, and public policy. He was focused on a broad problem that required linking many disparate areas of study.

He felt lucky. The economy had forced him out of farming, but he was beginning to "know more about agriculture than most farmers." Even so, he still did not know how he'd find a way to support himself and his family by applying GIS to agriculture.

He thought he'd have to work for a state or federal agency; in the mid-1980s the infrastructure in the private sector just wasn't there. The Department of Defense did not complete the GPS constellation until the mid-1990s, and even then the signals were scrambled and unreliable.

As a result, Waits took the most entrepreneurial of the options open to him once his Ph.D. was in hand. The National Aeronautics and Space Administration (NASA) was looking for an agricultural program manager to head up a commercialization project. In 1991 Waits loaded up the family again, this time for Mississippi. Though he was not a federal employee, his job was to take the technologies the government had been developing and spin out new companies that could capitalize on the new knowledge. By then Massey Ferguson's experimental yield monitors were making their way into commercial farm fields. But the monitors were rudimentary by today's standards, and the GIS data were cumbersome and expensive to use. There was progress, but it was tough going. Thinking back to those challenges, Waits asked, "Can you imagine asking a farmer to wait six hours to plant until the GPS signal came back on line?"

At NASA Waits assembled a team of people who had practical ideas about farming, along with technical modelers, coders, and engineers. Surprisingly, California growers were not interested in the applications Waits was studying. He found the greatest interest among progressive, high-value crop producers in the Midwest, particularly potato farmers in Wisconsin. In the early 1990s farmers didn't know about precision technology, remote sensing, or GIS. "Most producers don't want to mess with high tech stuff," Waits said. Producers needed to see the benefit, and they needed to be educated.

For that reason, among others, Waits briefly took a job as a professor at Oklahoma State University. After a couple years at the NASA commercialization center, Waits had learned that when GPS wasn't scrambled, it could have wide-ranging applications. While farmers could use airplane-based analog videography to make near real-time field decisions, digital images and GPS would open entirely new possibilities. For the first time Waits could see a way to make a living using what he had learned in the field he loved without working for a government agency. His NASA affiliation had provided connections and opportunities, but it also came with many constraints. Waits bided his time as a geography professor, something he had never imagined himself doing when he was a kid back on the farm. He taught courses in GIS and remote sensing, focusing on agricultural applications. His plan had been to start a company to apply GIS technologies to farming, and after a couple years he gave up his comfortable faculty job to put it all on the line. He started SST Software in 1994 and went at it full time in 1997.

In the early years of the business, collecting the data wasn't always easy, nor was making practical use of the information. When he started SST, personal computers weren't fast enough to speedily pull up maps. No one had iPhones or handheld GPS devices that could readily provide their location. The GPS that did exist was often unreliable. Web pages, clouds, and the other information technologies that we today take for granted did not exist. SST was able to take advantage of the new Internet technologies, but like so many other technology companies, it hit tough times in the years following the dot-com bust. Waits bought out a partner and friend from his farming years and regrouped.

Waits focused on progressive producers who were eager to learn and who had a strategic vision for the future. "We didn't begin with the profit motive in mind," Waits said. "Profit would come someday" if they did things right. He felt that GIS could provide farmers with a competitive advantage by optimizing the productive capacity of their fields. But not at the expense of profitability, as might be the case with other technologies. GIS provided the opportunity to improve the environment and profit. Waits insisted the technology would prove to be a win-win.

Subsequent research has shown that precision agriculture results in reduction of herbicide use by about 50 percent and typically reduces the application of nitrogen and runoff, cutting costs for farmers and improving environmental outcomes. As one group of academics puts it, "The concepts of precision agriculture (PA) and sustainability are inextricably linked. From the first time a global positioning system was used on agricultural equipment the potential for environmental benefits has been discussed." For wheat, research suggests that precision agriculture increases the efficiency of nitrogen use by 368 percent and reduces fertilizer use by 10 percent to 80 percent without reducing yield or grain quality. One researcher argues that "precision farmers probably meet more of the requirements for mitigating the effects of nitrate in ecosystems than do organic farmers."[10]

Of course, those environmental and economic benefits couldn't have been realized without innovators and entrepreneurs who developed technologies and took them to the farm. That's precisely what Waits did. Waits, and his company, developed around the idea that the key was to create a platform that could use the wealth of information that would be provided by the emerging precision technologies. Lots of people were thinking about remote sensing and precision agriculture, but "no one knew how to collect data or what to do with the data," he said. As yield monitors developed, software could take that data and turn it into a yield map—a multicolored image of the farm field that shows the areas that produce relatively more and less. A variety of companies today can provide these sorts of maps, but the real value comes when these yield data can be linked geographically with other information, like data from soil tests taken through the field, on the type of seed planted, fertilizers previously applied to the field, and so on. As Waits put it, "People can express data in a map, but the point isn't to make a map. The power is in the information that comes from combining maps. That's the difference between mapping and GIS." The idea was to create a spatial database management system.

Today SST Software is one of the leading agricultural suppliers of geographic decision support tools in the world. SST houses data on more than 100 million acres of farmland in twenty-three countries from Australia to Africa. At the heart of the operation is software that consists of relational databases that link information about the use of farm inputs to geographic identifiers and to site-specific information about soils, moisture, and much more. Some farmers can use SST's software directly themselves, but given the size and complexity of today's farms, the rapid pace

of technological change, and the expertise needed in entomol-
ogy, agronomy, and economics, many farmers rely on consultants
to help make management decisions. As a result, SST's biggest
clients are crop consulting companies like Crop Quest and Servi-
Tech and seed and chemical suppliers like Monsanto and Helena
Chemical. These companies often work with farmers to send to
SST information on soil samples, pesticide and fertilizer applications, yields, insect scouting reports, and seed varieties planted.
The companies use these data to make site-specific fertilizer or
planting recommendations. For example, based on his company's
agronomic models, an adviser might use SST software to identify which areas of a field should receive which kinds of fertilizer and in which amounts—a recommendation that can be sent
electronically to a variable rate spreader that communicates with
satellites to determine when and where to apply which mix of
fertilizers. Given the high cost of seed, new variable-rate planters are also coming on the market. A thumb drive loaded with a
recommendation from SST can be plugged into a planter, which
can plant two different corn hybrids at different seeding rates
and at different depths throughout the field.[11] SST doesn't make
recommendations; it provides the mechanism for translating an
agronomist's recommendation into an action plan.

Sitting through a tutorial on how to use the software provides some perspective on just how mindboggling are the choices
facing today's farmer. Once you identify a field on an aerial map
(think something like Google Earth) and pinpoint the precise
geographical location of the field and various management
zones, then you have to enter data. It's straightforward to upload

geographic information on production from a yield monitor, but sending in other information requires a human. Here the sense of scale emerges. Pick one of the nearly one hundred crops listed in a drop-down menu supported by the software, and then you have to make dozens of management decisions. For corn, you can pick seeds from major players like DeKalb or Pioneer or from smaller companies like AgriGold or ProHarvest. The software lists literally more than one hundred companies offering seed corn. After choosing a seed company, you have to select the variety or hybrid you planted. For the DeKalb brand of corn alone means picking from hundreds of varieties, with names like DK 307 or KD 234RR or 605F. What seeding rate was used, at what depth, on which parts of the field? And on and on. That's just for seed. What about fertilizer? Again, there are literally hundreds of possibilities, from generic 10–10–10 to dozens and dozens of branded varieties. Then the same can be done for insecticides and fungicides and a multitude of other possible supplements. Each field might have 100,000 possible items of information linked to it in a given year, and that doesn't begin to count the combinations of management decisions the farmer might have to make when mixing a particular type of seed with a particular soil type and a particular fungicide.

This not only illustrates the challenge of being a profitable commercial farmer in today's world, but it helps us begin to appreciate one reason why food is so abundant and affordable in the United States: the myriad technological innovations and concomitant options available to today's farmers. Companies like SST and the advisers who use their software are able to help farmers select from an overwhelming variety of options those that boost the bottom line while minimizing deleterious effects to the surrounding environment.

Big Data brings enormous possibilities. Want to know the highest-yielding soybean variety planted at different seeding rates in fine sandy loam within a specific four-county boundary in Indiana in 2011 that used a particular fertilizer? SST has the data to tell you. But it won't—at least not yet.

Just as Big Data has made us leery of the likes of Google and Amazon using our search and purchasing histories against us, some farmers are worried about large agribusinesses doing the same. SST does not own the data it houses but rather has licensing agreements with farmers or the service providers who enter data on behalf of farmers. Yet a real possibility exists that some of the companies that use SST software have better real-time information about crop yield conditions or harvests in the United States than does the federal government, raising the possibility that such information might be used to arbitrage commodity markets before the U.S. Department of Agriculture releases reports (think how Eddie Murphy's character in the movie *Trading Places* was able to profit by getting an early copy of the government's report on frozen orange juice concentrate).[12] These sorts of concerns recently led to an agreement on the use of farm data by major agribusinesses like Monsanto, DuPont, and Pioneer on the one hand and major farm organizations like the American Farm Bureau, National Farmers Union, and National Corn Growers Association on the other. The agreement says that the farmers own the data from their farms and that the agribusiness companies cannot use the farmers' data to speculate in commodities markets.[13] No doubt, new challenges and disagreements will arise as the technology develops, but, for the time being, it appears precision technology is here to stay.

Ironically, for a founder of a company that helped facilitate the collection and warehousing of Big Data, Waits spent much of his time in the early years of SST leery of the big guys. He was fearful "some big outfit would replicate what we were doing and take us over." With David Waits's son Matt Waits now at the helm, the future looks bright. In this dynamic, competitive environment, they may one day find themselves aligned with a larger company but only because they've created a product of incredible worth. As David sees it, SST was able to outlive a host of competitors with much deeper pockets, in part because the competitors couldn't see what made SST so valuable. One factor that gave SST a competitive advantage was that David and Matt, and the people they brought on board, knew agriculture. Yes, someone from Silicon Valley might know more about coding or cloud-based computing, but, as David asked me, "Can a guy from Stanford or Google understand why a farmer in Lockney, Texas, makes certain decisions? Understand the difference between Bray and Olsen extraction methods for phosphorus? They don't know what they don't know."

In this same way the advent of precision agriculture has perhaps had an unexpected benefit. When the Internet was first taking off in the mid-1990s, many prognosticators thought it would mean the "death of distance"—it would make location irrelevant and "eliminate geographical differences." But, as *The Economist* recently noted, reality has turned out differently, with mobile devices and other changes serving to "increase the importance of location" where "companies are increasingly treating the physical and virtual worlds as complements rather than alternatives." In many ways the Internet has allowed us to learn more about how we're different from others and has allowed us to find ways to satisfy our unique desires and whims. Likewise, precision technology has allowed farmers to know more than they ever have about their fields, and it is allowing them to be ever better stewards in unique ways suited for the geographies in which they reside.[14] The writer and farmer-activist Wendell Berry, fretting recently in an article for *The Atlantic* about the declining number of farmers, opined: "The great and characteristic problem of industrial agriculture is that it does not distinguish one place from another. In effect, it blinds its practitioners to where they are. It cannot, by definition, be adapted to local ecosystems, topographies, soils, economies, problems, and needs."[15] Yet in an era when local food advocates have encouraged consumers to reconnect with a sense of place, precision agriculture has allowed farmers to do just that.

7

Bovine in a Beaker

My grandfather was a walking Discovery channel.[1] Although he had never been to college, he had an encyclopedic knowledge of plants and geology. My grandparents' house was full of strange rocks, plants, and *National Geographic* magazines. Road trips with my grandfather took twice as long as they should have. Every few hours he'd make us get out of the VW van and look at some rock pile or give us a short lecture on the ecology of some wilderness area.

More than a few times during our teenage years my siblings and I rolled our eyes when a new hill appeared over the horizon and the car began to slow. Yet those lessons are now among the fondest memories of my childhood. My grandfather showed us some of the most remarkable discoveries during a trip to Terlingua, Texas—a remote, desert-like, mountain area within eyesight of Mexico. Best known for its annual chili cook-off, Terlingua is also home to some unusual plant and animal life.

My grandfather dug out of the ground a dead plant that looked like a dried-up, compacted tumbleweed. After we got it home, he gave it a bit of water and it miraculously unfurled and turned green within a few hours. The so-called resurrection plant adapts to arid climates by slowing its metabolism in periods of drought to such a point that it looks dead. Yet, with a little water, it springs back to life.

Perhaps even more interesting were the lizards that—to my surprise—would leave their tails in your hand as they scurried away into the desert. I was incredulous when told they would regrow their tail. We even found a few specimens that had grown a replacement after a run-in with a hawk, snake, or unruly kid. As it turns out, the lizard isn't alone. Salamanders can replace lost appendages, and many beachgoers know that starfish can replace arms and crabs can regrow lost claws.

Before I headed off to college, my grandfather suggested that I consider a career in biology and genetics, fields he saw teeming with exciting possibilities. Although his brother owned a restaurant, I'm not really sure what my grandfather would have thought of my choice to study food and economics. Nevertheless, I have little doubt about what he'd think of the scientists who are trying to harness regenerative powers akin to those possessed by lizards, starfish, and resurrection plants to produce a more abundant and sustainable food supply.

*   *   *

On August 5, 2013, Mark Post went out to grab a hamburger. This was no Big Mac from a drive-through. Post bit into his

$325,000 burger in front of an invitation-only crowd of jour-
nalists, chefs, and food enthusiasts in the heart of London. The
strangest part wasn't the cost or the crowd, but the meat. Post, a
professor of vascular physiology at Maastricht University in the
Netherlands who has spent time on faculty at Harvard Medical
School and is an expert on skeletal muscles, grew the burger himself. Not from a cow on his farm, mind you, but from a bovine
stem cell in a petri dish in his lab. Post's research, partially funded
by Sergey Brin, one of Google's cofounders, has the potential to
upend conventional wisdom about the environmental, animal
welfare, and health effects of meat eating.

Ironically, I met Post at a meeting of some of the world's largest hog producers.[2] Without any apparent trepidation, he took
the stage in front of several hundred hog farmers and meat processors and told them their business was causing undue animal
suffering and environmental harm.

Post's concerns are widely reflected in popular culture. For
example, the 2014 documentary film *Cowspiracy* argues that meat
eating is the single greatest environmental threat to the planet.[3]
Bill Maher, comedian and host of an HBO talk show bearing
his name, has written, "But when it comes to bad for the environment, nothing—literally—compares with eating meat. . . .
If you care about the planet, it's actually better to eat a salad in
a Hummer than a cheeseburger in a Prius." He's not alone in
his concern. The cookbook author and *New York Times* writer
Mark Bittman has asserted that our modern livestock production practices are leading to a "holocaust of a different kind."
The historian James McWilliams, in an op-ed piece in *The New*

York Times, perhaps summed up the prevailing view best when he wrote: "The industrial production of animal products is nasty business. From mad cow, E. coli and salmonella to soil erosion, manure runoff and pink slime, factory farming is the epitome of a broken food system." McWilliams argues that modern livestock agriculture is so damaging that the only moral solution is to give up eating meat entirely.[4]

These statements contain a bit of hyperbole, and I'll take issue with a few of the claims later, but for now it is important to note that these ideas are influential and have moved beyond popular discourse to affect public policy. For example, the 2015 Dietary Guidelines Advisory Committee released its recommendations to the secretaries of the U.S. Department of Health and Human Services and U.S. Department of Agriculture.[5] The preliminary dietary guidelines have dropped meat from the list of recommended foods, and the committee has, based on both heath and sustainability arguments, recommended a more plant-based diet and less meat consumption. The change in stance affects not only the information disseminated to the public but also what foods the national school lunch programs use and the flow of federal dollars to competing research priorities.

Post has taken a different tack—one that has sometimes put him at odds with vegetarian advocacy groups. Rather than selling the "tax meat" tote bags that People for the Ethical Treatment of Animals (PETA) uses to raise money[6] or cajoling and chastising people for eating meat, he recognizes that most people like to eat a good steak. Yes, some people live a healthy, vegetarian lifestyle for their entire lives. But that's not the norm. Only about

5 percent of the U.S. population is vegetarian or vegan, and 84 percent of the people who are vegetarians and vegans eventually go back to eating meat.[7] It seems that we are biologically wired to want to eat meat. That bite of filet mignon is a protein-packed, nutrient-dense morsel of tasty goodness. Anthropology shows that humans have probably eaten meat since the beginning, and some biologists believe that meat consumption, and the ability to increase nutrient content through cooking meat, played a role in increasing our brain size, making us into the species we are today.[8]

No matter the news on health and environmental outcomes, most people are still going to want to eat a burger. Indeed, one of the first things people at the lower end of the socioeconomic spectrum in developing countries want to do when they get a little more money in their pockets is to eat more meat.[9] Meat eating is positively correlated with income. Although meat is often more expensive than grains or veggies, when we can afford it, we want it.

Given that people want to eat meat, what can we do to make a difference? Why not try to make sure that the meat we do eat has as little environmental impact as possible?

One way to reduce the environmental impact of meat eating is to make livestock more productive. That is the route relentlessly pursued by the livestock industries. And it has worked. Beef production in the United States has a far lower carbon footprint than in other parts of the world, precisely because we use more intensive operations, and, despite the allure of grass-fed beef, feedlot operations for fattening cattle with grain have smaller carbon footprints. One study in the *Journal of Animal Science* calculated that from 1977 to 2007, the carbon emissions associated with producing a fixed quantity of beef in the United States fell 16 percent, water use fell 12 percent, and the amount of manure generated dropped 19 percent. Increased efficiencies have led to similar reductions in carbon emissions and resource use for hogs, poultry, and dairy. For example, one report from the pork industry suggests that the environmental impact and natural resources associated with hog production fell by 50 percent during the fifty years between 1959 and 2009.[10]

Unfortunately, some of these efficiency gains have been realized by using technologies that many consumers dislike. For example, use of growth promoters, such as added hormones, in beef cattle have significantly lowered carbon emissions, water use, and animal waste. But, despite clear evidence of the safety of these products, consumers are disturbed by the thought.[11] Likewise, the increased efficiency of pork production has occurred in part because hogs have been moved indoors and sows are housed, for a majority of their lives, in gestation stalls that, while preventing aggression and facilitating individualized feeding and care, prevent the mamma hogs from turning around.[12]

Post is imagining an altogether different sort of technological innovation, one that yields the bacon and burgers we love without the hog or cow. All animals have stem cells living in their muscles. These cells are capable of regenerating muscle cells for the animal. They're also capable of creating muscle cells outside the animal. The trick is to harvest stem cells from the muscle of a live cow or pig and turn them loose in the right environment in a lab. Stem cells proliferate quickly. A single stem cell can generate 100 trillion cells of meat. So, from a small number of donor animals (many millions fewer than currently exist to satisfy our appetites), we can ultimately get our burgers, chops, and chicken tenders without harm to the cows, pigs, or chickens. We no longer have to kill the geese to get the golden eggs.

Once in the lab and given a suitable environment, the stem cells naturally form into muscles. But the formation of muscle cells doesn't mean we have something ready to eat. We all have muscles, but they vary in shape and size. I've got a couple of scrawny biceps, whereas Arnold Schwarzenegger has guns. How did Schwarzenegger's arms get so big? No doubt genetics played a role, but he's also spent many, many more hours in the gym than I. Building protein and muscle size requires exercise. Post and his team attach the muscle cells to each other and to a proverbial petri dish in a way that creates exercise-like tension. The result is muscle fiber (each fiber contains about 1.5 million cells). Once these muscle fibers develop, about ten thousand are harvested to create a burger.

This isn't a soybean patty crafted to look like a hamburger. It is a real meat burger produced by the same cells doing the work in a real cow. Still, when Post talked about his new burger, he said it was a bit dry. That shouldn't be too surprising because it was 100 percent lean and had no fat. To really mimic the taste of a burger, fat and connective tissue are needed. These fats could come from vegetable sources (like canola or corn oil) or Post might just grow it too. He's now got a few petri dishes of animal fat growing to try to add a little flavor to his next lab-grown lunch. Fat stem cells from livestock can be used in much the same way as muscle stem cells, although they're a bit trickier to deal with because they're more flexible than muscle stem cells and must be coaxed into producing only desirable kinds of fat.

Perhaps surprisingly, not all vegetarians and animal advocacy groups are supporters of Post's work. For one, they don't like it that stem cells must be extracted from a live animal. But if the choice is between the emissions of carbon by a few thousand animals and many millions of animals that leave waste in waterways and gobble up valuable resources that we humans might use, the answer seems clear. But these folks also point out that lab-grown meat isn't a free lunch. The stem cells have to eat something to grow. Right now the cells grow in a medium that relies on animal-based serum. However, Post and others are devising non-animal–based feed stock for the muscles.

More broadly, this line of argument—that meat production (inside the lab or out) is wasteful because it requires feed resources that humans might use—is misplaced. To see this, it is useful to consider a thought experiment—an imaginary story that might help us get to the bottom of things.

Imagine a biologist on an excursion to the Amazon to look for new plant species. She comes across a grass she's never before seen and brings it home to her lab. She finds that the grass grows exceedingly well in greenhouses with the right fertilizer and soil, and she immediately moves to field trials. She also notices that the grass produces a seed that is durable, storable, and extraordinarily calorie dense. The scientist immediately recognizes the potential for the newly discovered plant to meet the dietary demands of a growing world population.

But there is a problem. Lab analysis reveals that the seeds are, alas, toxic to humans. Despite the setback, the scientist doesn't give up. She toils away year after year until she creates a machine that can convert the seeds into a food that is not only safe for humans to consume but that is incredibly delicious to eat. There are a few downsides. For every five calories that go into the machine, only one comes out. In addition, the machine uses water, runs on electricity, burns fossil fuels, and creates carbon emissions.

Should the scientist be condemned for her work? Or hailed as a genius for finding a plant that can inexpensively produce calories, and then creating a machine that can turn those calories into something people really want to eat? Maybe another way to think about it is to ask whether the scientist's new food can compete against other foods in the marketplace—despite its inefficiencies (which will make the price higher than it otherwise would be). Are consumers willing to pay the higher price for this new food?

Now let's call the new grass corn and the new machine cow.

This thought experiment is useful in thinking about the argument that corn is wasted in the process of feeding animals (or growing meat in a lab).[13] Yet the idea that animal food is wasted is a common view. For example, one set of authors wrote in the journal *Science:*

> Although crops used for animal feed ultimately produce human food in the form of meat and dairy products, they do so with a substantial loss of caloric efficiency. If current crop production used for animal feed and other nonfood uses (including biofuels) were targeted for direct consumption, ~70%

more calories would become available, potentially providing enough calories to meet the basic needs of an additional 4 billion people. The human-edible crop calories that do not end up in the food system are referred to as the "diet gap."[14]

The argument isn't as convincing as it might first appear. Few people really want to eat the calories that come directly from corn or other common animal feeds like soybeans. Unlike my hypothetical example, corn is not toxic to humans (although some of the grasses cows eat really are inedible for humans), but most people don't want to eat field corn.

So if we don't want to eat the stuff directly, why do we grow so much corn and soy? These plants are incredibly efficient producers of calories and protein. Stated differently, these crops (or grasses, if you will) allow us to produce an inexpensive, bountiful supply of calories in a form that is storable and easily transported.

The assumption seems to be either that the diet gap will be closed by convincing people to eat the calories in corn and soy directly or to consume other tasty crops that can be grown instead of corn and soy but can be grown as widely and produce calories as efficiently as they can. Aside from maybe rice or wheat (which also require some processing to become edible), the second assumption is almost certainly false. Current consumption patterns suggest we should be skeptical that large numbers of people will voluntarily consume substantial calories directly from corn or soy.

Typically we take our relatively un-tasty corn and soy and plug them into our machine (the cow or pig or chicken or, in Post's case, the petri dish) to get a form of food we want to eat.

Yes, it seems at first to be inefficient, but the key is to realize that the original calories from corn and soy were not in a form most humans find desirable. As far as the human palate is concerned, not all calories are created equal; we care a great deal about the form in which the calories are delivered to us.

The grass-machine analogy also helps make clear that comparing the calorie and carbon footprint of corn directly with the cow is probably a mistake. Only a fraction of the world's caloric consumption comes from directly eating raw corn or soybean seeds. It takes energy to convert these seeds into an edible form— either through food processing or through animal feeding. So what we need to compare is beef and other processed foods. Otherwise we're comparing apples and oranges (or, in this case, corn and beef).

The more relevant question is whether lab-grown meat uses more or less corn, and creates more or fewer environmental problems, than does animal-grown meat. Given current technology, the answer isn't yet clear. Post projects that if his current system were scaled up, he could produce a pound of lab-grown meat for about $30. Some people in niche markets are willing to pay that much for a burger, but it's far higher than today's prices. In 2014 (a record-high year for beef prices) the average price of hamburger in the United States was only about $4 a pound.[15] There are no doubt costs associated with beef production that are not reflected in the retail price (e.g., the impacts on the environment or human health), but it's hard to imagine these extra costs are 650 percent more than the current price of a burger. Post has a ways to go if he's to compete with the cow.

But he's not giving up. And he has competitors like the U.S.-based firms Modern Meadow and Impossible Foods (the latter of which has received financial backing from Bill Gates).[16] As Post invests more time in his project, and the technology improves and science develops, the price of lab-grown meat will fall. And, if Post and his colleagues can engineer a burger with a healthier fat profile in a way that is friendlier to animals, many of us might even be willing to pay a premium for it.[17]

Post has a few more ideas that might entice adventuresome eaters. If stem cells from cows can grow a hamburger, why not take stem cells from a rhino and create a rhinoburger? Or mix a few stem cells from a giraffe or rabbit to create a truly unique delicacy? More seriously, Post has in mind a tabletop meat grower that may one day grace the counters of our kitchens. We might one day have a homegrown burger with our home-brewed lager.

When it was finally my turn to address the meeting of pork producers, I could see that a few attendees were reeling from Post's remarks. I told them the same thing I tell my cattle-feeding buddies: if your business is threatened by the prospect of lab-grown meat, now's the time to invest.

8

Sustainable Farming

Like so many organizations, a few years ago my university decided to kick off a sustainability initiative. The effort encompassed the entire campus and involved inquiries into what we teach and when the heating and air should shut off at night.[1] One of my colleagues served on a university-wide curriculum committee that discussed sustainability in the classroom. After one meeting I asked him, "So, how'd it go?" He replied, "Well, I finally figured out the definition of *sustainability*." "What's that?" I asked. He said, "It's whatever actions will result in *my* department getting more resources."

One of the problems with the concept of sustainability is that no one really knows what it means. Sure, there are academic discourses that use Venn diagrams to try to nail down the elusive term. Some environmentalists believe sustainability can be achieved only by scaling back economic growth and learning to live with less. Others believe growth is the only way we can be sustainable. I once heard an organic farmer say, "The first rule of sustainability is I've got to make enough money this year to do it again next year or game over."[2] Depending on whom you ask, sustainability could mean everything from economic efficiency to environmental justice to organic, locally grown free-range chickens. And, of course, a lot of skeptics think that big business has co-opted and greenwashed the term. It's now common for executive suites to house chief sustainability officers (CSOs) next to CEOs and CFOs.

Yet underlying all the cynicism is something we can all probably agree is really important. Will our children and grandchildren be able to enjoy the same standard of living that we've come to expect? Prudently and judiciously using our resources in a way that we leave enough for retirement and the next generations is a worthy goal. But it's really hard to think that far into the future. Research suggests that we can't imagine the future more than fifteen to twenty years out.[3] Thus to gain insights into the future, and our ability to sustain our lifestyle, looking backward might be useful. History, rather than imagination, might give us a more reliable glimpse of the future.

Just two years after Congress created the Oklahoma Territory out of what were Indian lands, a young researcher by the name of Alexander C. Magruder started a scientific experiment on sustainability that continues to this day.[4] The year was 1892, and the Oklahoma Territory was in the midst of a population surge, thanks to the land runs and westward migration. The twenty-four-year-old Magruder was lured to Oklahoma Territory by an annual salary of $1,500.

With a B.S. in agricultural science from Mississippi Agricultural College, where his father was an English professor, and after further training in Germany and West Virginia, Magruder became the first professor of agriculture at Oklahoma A&M College (what is now Oklahoma State University). The young Magruder was a sharply dressed, well-mannered man who was described as aristocratic. He was also ambitious. Within a few short years Magruder purchased a new John Deere tractor; ran feeding trials for livestock; studied irrigation and crop rotation techniques; planted corn, cotton, cowpeas, clover, flax, oats, sorghum, and wheat (some from seeds he brought back from Europe); promoted short courses for farmers; taught classes; and created gold medals to hand out to prize students. During all this he somehow managed to earn a master's degree from his undergraduate alma mater in 1894.

Magruder's legacy stems from an experiment he decided to run in late 1892. On a plot donated to the college by a local family, Magruder plowed up virgin prairie soil to explore what would happen if the land was "sown in wheat year after year without the addition of any fertilizing material." In short, Magruder was interested in the sustainability of farming practices that relied on no outside sources of nutrients. He wrote in 1892, "No fertilizers, either commercial or home-made, were used. It is our object to get at the natural value or strength of the soil that we may compare present yields with those of the future when barn-yarding

and green manuring will have been practiced."[5] This was before
the German chemists Fritz Haber and Carl Bosch figured out
how to extract nitrogen fertilizer from the air. Mineral fertilizers were in short supply and were not widely used. The main
fertilizer available to farmers was what it had been for centuries:
animal manure.

Today we take for granted the easy accessibility of fertilizer.
We head down to the grocery or hardware store and buy a box
of Miracle-Gro or a bag of Scotts Turf Builder. But until about
three hundred years ago, there were only a few ways to boost crop
yields. One option was to plow virgin land to gain access to nutrients held in the soil. But after a few years those nutrients were
typically depleted. Rotating in a nitrogen-fixing crop, like beans,
could help. But the main option was to add compost and manure.
So important was manure to crop yields that the ancient Romans
elevated excrement to deity status by paying homage to Stercutius,
the god of manure. But there was never enough to go around.

In a very real sense the world's population was limited by
the availability of soil nutrients, particularly nitrogen. Then nitrogen fertilizer was accidentally discovered in South America.
As the story goes, some traders were crossing a desert near the
Andes Mountains in the 1700s when they were shocked to find
the ground beneath them was set ablaze by their camp fire. Fear-
ing the work of the devil, some of the soil was brought to a priest
who, thinking it was merely saltpeter of insufficient strength for
gunpowder, threw it out. However, a few weeks later, he noticed
that the plants in the area where the saltpeter was discarded grew
taller, greener, and more vibrant than the others around it. The

revelation was passed on to a British naval officer, who brought the news back to Europe. The find eventually led countries to scurry into geopolitical battles in the 1800s to shore up mineral deposits, but those sources, along with the Peruvian guano mines, were quickly becoming tapped out. Nitrogen remained scarce until early in the twentieth century, when Haber and Bosch—both of whom were awarded Nobel Prizes—discovered an economical way of pulling nitrogen from the air in the form of ammonia. And we haven't looked back. In a fascinating book, *The Alchemy of Air,* Thomas Hager writes,

> As a species we long ago passed the natural ability of the planet to support us with food. Even using the best organic farming practices available, even cutting back our diets to minimal, vegetarian levels, only about four billion of us could live on what the earth and traditional farming supply. Yet we now number more than six billion, and growing, and around the world we are eating more calories on average than people did in [the late 1800s].[6]

How did this happen? Haber and Bosch.

While Haber was just beginning to work on the nitrogen problem in his lab in Germany, Magruder was in Oklahoma, working on how to increase nitrogen availability. He advocated the planting of nitrogen-fixing cowpeas in 1894, arguing that "it would be well if each Oklahoma farmer could . . . instead of letting our lands be exhausted by the one-crop system . . . begin now to grow crops that will maintain the soil's natural fertility."[7]

Magruder's original experiment with the virgin plot was designed to see just how quickly soil fertility declined and whether the addition of manure or other fertilizers could help offset loss of soil nutrients to provide sustainable yields.

Magruder soon ran into a patch of bad luck. In 1894 his fiancé, Bessie Duncan, became ill after a twenty-mile trip in a rain-soaked buggy. Worried about her prognosis, Duncan requested that the two be married. She died only six days later. Then, in 1895, he ran afoul of campus politics. The members of the university's Board of Regents were political appointees. This was a time before tenure, and professors could be dismissed for any reason, including lack of party loyalty or holding politically incorrect views. For example, about 250 miles north at the Kansas Agricultural College, when the Populists and Democrats were elected, newly appointed college leadership canceled all faculty contracts in 1896 in retaliation for the faculty's perceived sympathy for the Republican Party (though many professors were later rehired). Magruder ran into trouble with the regents in a dispute about the sale of cattle that had been a part of a feeding trial. He returned from vacation in 1895 to find himself without a job. Magruder was run out of town before Oklahoma even became a state. Fortunately, his research project had a longer tenure.

Magruder returned to school at Tulane University and became a doctor; he successfully practiced medicine (and kept a herd of registered pigs) in Colorado until his death in 1924. In Oklahoma the work Magruder started has continued for more than a century. For more than 120 years, the plot has been planted with wheat every year, without being allowed to lie fallow, never has been amended with fertilizer, crop rotation, or the planting of a cover crop. The plot that now bears his name is the longest-running soil fertility study of wheat on the Great Plains. Only the University of Missouri has soil fertility research on wheat that has been running longer; it was established in 1888. The University of Illinois has the world's longest experiment involving continuous corn, begun in 1876 by George Morrow (who, incidentally, was appointed president of Oklahoma A&M in 1895). Sustainability is related to the abiding effects of management practices and decisions, and precisely these sorts of long-term research projects provide insights into the dynamics of soil health.

Century-old research projects required not only foresight at the outset but persistence and creativity to keep them going. On more than one occasion the Magruder plots faced existential threats. After fighting off earlier campus expansion projects, agronomists lost the battle when school administrators wanted the prime real estate to accommodate the swelling enrollment of veterans returned from World War II. But the researcher in charge of the plots at the time, Horace Harper, triumphed nonetheless. Having conceded the geographic location to a new dorm, he sought to save the soil.

Harper had been on the faculty at Oklahoma A&M during the Dustbowl, which left Harper with deep concerns about erosion and soil fertility, and he fought hard to save Magruder's project. In the summer of 1947 Harper salvaged the experiment by securing a contract to move five hundred tons of soil, the top fifteen inches from each plot, from the original location to a new location about a mile away. By the fall of 1947 the dirt had been moved, the plots were again replanted in wheat, and the experiment continued. When plans to expand the campus were renewed in the 1970s, the rumor was that the new location of the Magruder plots was under consideration for new athletic facilities. Seeking to preempt disruptions to the research, the agronomists in charge of it got the Magruder plots named a historic landmark, and they were added to the National Register of Historic Places by the Department of the Interior in 1979.

What have we learned from the century-old soil fertility research? To answer that question, I walked down four flights of stairs to talk to my colleague Bill Raun, a Regents Professor in the Department of Plant and Soil Sciences at Oklahoma State University. Raun studies soil science and precision agriculture, and he has been the researcher in charge of the Magruder plots for more than twenty years. Raun exudes passion for his job, which for him is something of a calling. I couldn't help but notice the images of Norman Borlaug and Mother Teresa, two Nobel Peace Prize winners, serving as his computer's desktop background. He talked about his six years in Mexico and Guatemala as an agronomist for CIMMYT—which is the Spanish acronym for the nonprofit International Maize and Wheat Improvement Center. Raun said the near-constant dysentery he suffered while running programs in nine Central and South American countries eventually wore him out and led him back to the States and ultimately the Magruder plots.

I asked Raun to identify the most important lesson the Magruder plots have taught agronomists after more than a century of research. He said, "After a hundred years, it's still producing almost twenty bushels of wheat. It's amazing!" (for some perspective, the average yield in Oklahoma from 2010 to 2015 was 27.6 bushels per acre).[8] Yields vary from year to year because of weather, among other factors, but, surprisingly, the overall trend in yields on the unfertilized plot is, if anything, positive. The continued productivity of the land after a hundred years is, according to the title of one of Raun's publications, a puzzle. After all, his research has shown that the organic matter in the soil has fallen from about 4 percent when the native prairie grass was originally plowed to just more than 1 percent today. Raun said the loss of organic matter has slowed and that the larger initial losses occurred soon after the plot was first plowed as some of the nutrients in the soil oxidized.

Despite demonstrated losses of soil nutrients over time, wheat yields haven't deteriorated. Why? Raun speculates the reason is genetic improvement. Farm practices at the plot have remained relatively constant over time, but about twenty successively improving varieties of wheat have been planted on the plot. Thus, in terms of crop yields, gains in genetic improvement have more than compensated for the losses in soil nutrients.

The good news, however, is that the farmer does not have to live with lost soil nutrients. Magruder began by planting only a "check" plot with no added fertilizer, but shortly thereafter researchers also included plots fertilized with manure to determine whether the addition of animal waste could compensate for lost soil nutrients. Then, beginning in the 1930s, researchers added even more treatments to determine how added macronutrients like nitrogen, phosphorous, lime, and potassium affected yields.

In the early years adding extra nitrogen didn't appear to provide much of a yield boost, suggesting that phosphorous was the key nutrient limiting the ability of the crop from heartier growth. The prairie soil was a rich source of nitrogen, and it took seventy years of "mining" before the lack of nitrogen would start to limit plant growth. Since the 1960s treatments that add nitrogen have steadily outperformed the application of phosphorous alone.

The key message from the 120-year-old experiment on the Magruder plots is that, so long as genetics continue to improve, and especially if manure or nitrogen can be added back to the soil, wheat yields on the Great Plains are not only sustainable but can experience continued growth.

Back in the 1890s, Magruder's original check plot yielded fewer than thirteen bushels per acre. Remarkably, in the first decade of this century, that same check plot averaged fifteen to sixteen bushels per acre. Plots that have received only manure treatments for more than 125 years are now yielding more than thirty bushels per acre, and the plots receiving nitrogen, phosphorous, lime, and potassium today routinely yield more than thirty-five bushels per acre. Even as far back as 1975, the researcher overseeing the plots concluded: "The lesson is that we can grow continuous wheat without harm to the soil and that fertilization is needed to replenish what the crop takes."[9] A few years ago Raun and his colleagues wrote, "While continuous wheat without rotation is not recommended, this 114-yr study documents the feasibility."[10]

The experiments at the Magruder plots show the amount of food produced on a given plot can more than *double* if nutrients are added. But how can the farmer do that in a responsible way? Raun has focused much of his research on answering that question—by developing technologies to apply fertilizers more efficiently. Ideally, the farmer should apply enough nitrogen that a plant can achieve maximum production potential but not so much that excess nutrients leech into waterways. A plant's nutrient needs depend on previous uses of the soil and they vary year to year and through the year based on rain, temperature, and other factors. As I have discussed, excess nitrogen can result in some nasty environmental consequences. Raun collaborated with the agricultural engineers John Solie and Marvin Stone to develop a sensor that can determine, in real time, the nitrogen needs of the crop at any point during the growing season.

The resulting GreenSeeker technology is a handheld device that senses the color of the plants' leaves and, along with other information—such as the date the crop was planted—provides a recommendation of how much nitrogen to apply to satisfy the plants' needs. The sensors can also be placed atop a tractor or fertilizer applicator so that, as a farmer drives through a corn or wheat field, the amount of nitrogen applied changes in response to what the GreenSeeker is "seeing." Today several companies make the GreenSeeker and related technologies commercially available to farmers all over the world. Adoption has been limited by the cost of the technology, lack of precision, and the low cost of alternatives, like simply applying a uniform rate of nitrogen throughout portions of the field before planting.

But Raun is optimistic and undeterred. He is concerned about excess application of nitrogen and the current high levels of fertilizer recommended in many parts of the country. As he sees it, if Congress and regulators aren't going to regulate the amount of nitrogen that can be applied, then farmers ought to use technology to get recommendations that more precisely meet the needs of each farm in a way that doesn't cause undue environmental harm. Raun's current research is taking that idea to its extreme. He's found, even in seemingly homogeneous corn fields in the upper Midwest, an incredible variability in yields from plant to plant. Even in the rich soils of Iowa, one cornstalk can yield the equivalent to 150 bushels per acre, while the one standing right next to it yields almost nothing. Some stalks of corn have a yield tantamount to 330 bushels per acre, and less than two feet away are stalks with a yield that would work out to only thirty bushels per acre.[11] Raun's newest experiments are with tractors that apply fertilizer on a *plant by plant* basis, applying nitrogen only when and in the amount that each plant needs it. Precision agriculture can't get much more precise than that.

Despite all this high-tech wizardry, Raun has been unable to ignore the decidedly low-tech forms of agriculture he saw in the developing world. He estimates that 60 percent of maize in the developing world is planted *by hand*. That's more than seventy-one million acres on Earth where poor, often subsistence, farmers use long sticks to poke a hole in the ground and drop in three or four kernels of corn before moving a foot or so and repeating the process again—and again and again. This is imprecise, back-breaking work, and potentially deadly.

According to Raun, many of these rural farmers have access to high-quality hybrid seeds, but the seeds have been pretreated with fungicides and insecticides. The treatments protect the vulnerable seedlings from insects and disease, but chemically treated seeds weren't meant to be routinely handled by the farmer. In the United States that's not much of a problem as bags of seed are dumped into large mechanical planters that do the work, but health risks abound in the developing world, where each seed is planted by hand. Raun again teamed up with engineers to create what he calls a GreenSeeder—a handheld device that can be loaded with seed and reliably deliver a single seed with each poke in the ground. Raun optimistically estimates the device could boost yields by 25 percent, resulting in $2 billion of extra revenue to the developing world if farmers abandoned their wooden sticks in favor of his mechanical poles.[12]

While Raun has focused his attention on the soil, others are focusing on the plant. And an important plant it is. Despite the gluten-free fad, wheat remains one of the most widely grown crops in the world, accounting for an estimated 20 percent of the calories consumed worldwide. Wheat is one of the oldest domesticated plants, and from its origins in the Fertile Crescent it became the staple crop of civilizations in Africa, the Middle East, and Europe. The largest producers of wheat today are also the largest population centers—China and India. No matter where you live, it probably wouldn't be too hard to find wheat, whether in beer, croissants, dim sum, hot dog buns, pita, tagliatelle, or tortillas.

Humans produce a lot of wheat and need more of it. According to a 2014 press release from the United Nations, "over the next 35 years farmers will need to increase the annual production of maize, rice and wheat to 3 billion tonnes," but they need to do it sustainably, "with less water, fossil fuel and agrochemicals, on farmland that has been widely degraded by decades of intensive crop production."[13] How are farmers to accomplish that seemingly impossible task? The Magruder plots help provide an answer. Even on land that had been intensively farmed with no agrochemical fertilizers for more than 120 years, it was still possible to get more food through genetic improvement and by planting better varieties of wheat.

Even Magruder recognized the importance of improving wheat genetics, writing in 1892, "If we can introduce a variety of wheat or oats which yields one bushel more to the acre than the one we now have, we will have done more good thereby than the equivalent of usual legislative appropriation for a series of years." Breeding wheat was not only about increasing yields but also about reducing downside risk, as Magruder recognized: "If we can save the farmers only one bushel per acre by preventing a decrease of the present yield, we have rendered here equally as much assistance."[14]

The continued improvement in wheat varieties comes from the work of geneticists and breeders like my colleague Brett Carver, Regents Professor in the Department of Plant and Soil Science. Just as the wheat harvest was beginning in early summer, I met up with the incredibly busy Carver at the university's Agronomy Research Station, which is home to the Magruder plots, along with hundreds of acres of other research.[15]

Carver took me out to the middle of an unusual-looking wheat field. The feeling of awe and beauty that comes when you look out at amber waves of grain arises, in part, from the many acting as one: each stalk and head of grain is about the same height and size, and the group moves in unison with the wind. But this wasn't that type of field. Carver's field looked a bit like a bad hair day. It was chaotic. Some stalks of wheat were almost up to my waist, others were only a bit taller than ankle height. Some stalks were golden yellow, others were darker brown. Some spikes scrawny, others fat. Long bristles protruded from most of the plants' heads, but some had no bristles.

Carver's goal is to create a new wheat variety. When a farmer buys seed of a known variety, they can be sure that the resulting plants will display a certain set of characteristics. In this way, a seed variety is like a brand. One Ford Taurus is similar to another Ford Taurus. If you buy a Taurus, you generally know what you're going to get. Buying a Porsche 911 will provide a whole different set of characteristics, but even still most Porsche 911's of the same year model are fairly similar because they're the same brand. Auto manufacturers design a new brand (or model) by making a prototype they think will satisfy their buyers' demands for speed, comfort, gas mileage, and more. Once they settle on a potentially successful design, a factory replicates the car over and over to create a new auto line (or we could call it a car variety).

Carver, by contrast, can't make a new wheat prototype from scratch. He has to go looking for a needle in a haystack that will provide the characteristics farmers, wheat millers, and consumers desire. Standing in the middle of the proverbial haystack he planted, Carver said, "There are sixty-six thousand different strains out here. I'll pick one of them, and it will ultimately be grown on millions of acres. It's a big responsibility." Carver developed all the top four varieties of wheat planted in the state of Oklahoma——Duster, Endurance, Gallagher, and Ruby Lee—where farmers planted more than five million acres of wheat in 2015.[16] Back in the greenhouse, Carver and his team start with about a thousand different parents and create crosses (or "kids") that they ultimately plant, making them some of the sixty-six thousand candidates for a new variety. About a hundred have almost no chance of making it in a final variety, but Carver said he plants them anyway because they might reveal a trait that will be useful in a future cross.

Wheat breeding is no simple matter. There's a lot of science involved, but there's an art to it as well. Carver has to stay abreast of the latest developments in molecular genetics to identify genes with useful properties, but he also has to watch and see what actually works in the field and make judgment calls through the growing season. Carver said there's a big misperception that "I'm out here just trying to increase yields." Carver has to consider about fifty traits when he is developing a new variety, including everything from how the wheat will withstand grazing by cattle (a common practice for Oklahoma farmers) to flour quality for millers to whether the wheat can survive myriad pests and diseases.

Yes, farmers want to produce more grain, but, as Carver put it, "There are a lot of different ways to get from here to Chicago," and he works on charting a path that Mother Nature will allow. In fact, one of Carver's biggest challenges is just trying to stay ahead of the problems that Mother Nature throws at him. A wheat crop can be devastated by fungi, viruses, or other diseases, and just when Carver thinks he's got one of them licked, the disease will mutate and take over again.

One of the biggest challenges Carver was facing in 2015 was a fungal disease called stripe rust. He said it was first noticed in Oklahoma in 2000, when it was probably tracked into the country on a worker's shoes. He reached over and ran his hand through a stalk of wheat to show me the damaging effects of the rust. As he said, "There's just nothing there." He compared it to the sturdy stalks next to it. The stalks of the plant infected with stripe rust were like thinning, wispy hairs on the forehead of a balding, middle-aged man, whereas healthy plants had thick, sturdy stalks. Farmers can apply fungicides to fight the rust, but it would be less costly and more environmentally friendly if Carver could identify those plants that possess the genetics that make them less susceptible to the disease, and eventually create a new variety that contained the genes. Or, as Carver put it, "We need a new vaccine"—one that's bred into the plants.

The trouble is that developing new plant varieties is a long process, typically taking ten to twelve years. Carver likened it to raising a child through to high school graduation. Once Carver identifies a handful of potential bright spots among the sixty-six thousand kids, he'll take them back into the greenhouse for the next growing season and subject the offspring to all kinds of tests and stresses to see what might be worth developing further. The ones that survive testing come out ready for middle school and junior high. Carver showed me some plots of "seventh graders and eighth graders," and we drove around, finally making our way to the seniors, which are about ready for graduation—release to farmers.

During successive generations Carver is removing all the undesirable genetic diversity so that he ends up with a pure line that will be introduced as a variety. Farmers don't want fields like the bad hair day. They need all the seeds they plant to come out about the same time, to respond in similar ways to fertilizer and grazing and to be ready for harvest at the same time. Getting this level of predictability requires years of careful selection. In short, wheat breeding works something like a genetic funnel. Start with a large pool of diverse genetics and then, year after year, winnow down to a small set that represents something farmers will actually want to grow.

It's here that new scientific and technological developments are making Carver's job easier. If he can reduce the time it takes to develop a new wheat variety, the sooner farmers have the "vaccine" and can cut back on fungicides without suffering devastating losses. One technique he uses is molecular breeding. Carver uses genetic tests to identify which genes the plants possess before he creates crosses, so that he ensures the kids will have the traits he's after. That's where having access to large amounts of germplasm (essentially seeds) with a wide array of genetic diversity is really useful. In fact, as Carver sees it, one of his main jobs is collecting and maintaining a huge database of germplasm (a seed bank, if you will) so that when a new threat arises, he can search through and find varieties that have the right genes.

Another new technique being used today by almost all wheat breeders is double haploid technology. To understand a bit how this works, think back to high school biology, where you learned that humans have twenty-three pairs of chromosomes. The pairs arise because we get half our genes from our mom and half from our dad. Each set of genes is called a haploid set, whether it's from your mother or father. Put two haploids together and you've got a full diploid set. Wheat, like any plant, puts those sets together naturally, but this takes a long time when going from a cross of two parents to a uniform pure line of kids suitable for growing. In fact, it takes about seven or eight years, not counting several more years of testing the offspring in field trials. Breeders needed a way to create genetic diversity through crosses yet turn that diversity into uniform pure lines in a much faster manner. If a pure line variety is genetically uniform, it will look essentially the same this year as it did last year. Breeders have figured out that if they can take one haploid set of chromosomes from a cross and copy it (or double it creating a double haploid), they will have a pure line almost instantly. The beauty of this technique is that the biological process of genetic reassortment (what makes each of us different from our siblings) remains unchanged from its natural state. This process is simply fast-forwarded.

There are different ways of creating double haploids. One method is by crossing wheat with corn. Wheat thinks it's being pollinated by another wheat plant and sends over half its genes (the haploids), but when they can't match up with the corn's genes, you're left with the single haploids that you wanted in the first place. Carver said that when double haploid technology is combined with molecular breeding, a new variety can be created in half the time and with better accuracy than field breeding.

That would be like creating a new grade school and curricula that graduate sixth graders who could score better on the ACT than today's high school seniors. Newer varieties in half the time mean Carver can double the genetic gains in yields, drought resistance, and disease resistance, among other traits of interest to farmers, millers, and consumers. And he'll be able to respond to whatever challenges arise in the future. "God knows what we'll be dealing with in ten years," Carver said, "but it will almost certainly involve handling drought and heat."

If all this sounds a little spooky or unnatural, taking a closer look at wheat's DNA might be useful. It's a fascinating story that lays bare some of our unjustified angst about unnatural food. Even though wheat has been around since the dawn of civilization, it is actually a product of biotechnology. But, as Carver said, "Man didn't do it. . . . God did it or nature did it, but it wasn't man." He added, "If I tried to do this today, I'd be labeled a mad scientist who's creating some sort of evil genetically modified food."

The history of wheat can be found in its DNA. Unlike humans, wheat does not have one father and mother but three fathers and three mothers. Rather than a single pairing of genes, which is what occurs in humans (a diploid), wheat has three sets of chromosomes, and each set exists as a pair—something called a hexaploid. This somewhat strange state of affairs came about when one species mated with another, and then it happened yet again. Carver explained that about 300,000 years ago one grassy weed species crossed with another—a spiky, unruly-looking plant that eventually led to the plant we call emmer. Then, about ten thousand years ago, this crossbreed mated yet again, with another grassy species, one of the many goatgrasses. The result is our modern wheat used for making bread; modern wheat carries DNA from all three of its distant relatives. Thus two major events in agricultural history caused cross-species gene transfers to create the plant we now call wheat. In fact, if we go back even further, some geneticists think, rice, corn, and wheat all share a common ancestor from fifty-five to seventy-five million years ago. All this makes Carver's job more complex. Whereas humans have an estimated 20,000 to 25,000 genes, wheat has 164,000 to 334,000 genes.[17]

All this talk of genetic change in wheat runs up against the claims in popular books like *Wheat Belly* and *Grain Brain* that have sparked the gluten-free fad diets. These books claim that many of our modern health problems, everything from diabetes to obesity to acne to arthritis, can be traced to eating wheat in general and the wheat-protein gluten in particular. Moreover, these authors suggest that modern wheat varieties have been manipulated in ways that have created a toxicity that wasn't present in older wheat varieties.

Asking Carver about some of the claims in these books brought him nearly to exasperation. He said, "Gluten is gluten. It's been there forever." As have the components of the gluten protein—gliadin and glutenin. So there's no reason to think that gluten intolerance is a widespread problem among humans. The durum wheat used to make pasta has a different set of genetics (it is only a tetraploid), yet it still produces gluten. Heritage varieties of wheat like einkorn and emmer wheat (which represent variants of the parents or kids of the crosses that occurred 300,000 and 10,000 years ago to produce modern wheat) contain gluten. Our bodies don't know whether gluten came from einkorn, durum, or bread because it's the same protein in every case. And we've been eating it since the beginning. Moreover, Carver said, he and other colleagues have gone back and looked at the nutrient content of wheat over the years (recall that one of Carver's main jobs is to keep a sort of seed bank to use in his breeding program), and there simply hasn't been much change in protein content or in content of other vitamins or minerals over time that would lead to gluten intolerance.

It is instructive to recognize that the gluten-free diet fad is a decidedly American (and to some extent Canadian and western European) phenomenon even though wheat is widely consumed everywhere in the world without reports of the wide-ranging maladies gluten supposedly causes (beyond celiac disease, which affects about 2 percent of the population). The United States exports about half the wheat it produces and is a major supplier to places like Europe, Asia, and Africa.[18] It's probably true that many of us would do well to cut back on the consumption of carbohydrates, but pinning all our dietary problems on a single wheat protein is trying to solve a complex, multifactor dietary problem with a single rule. In fact, the authors of one of the original scientific studies that helped foment concern about gluten sensitivity later went back and took a more rigorous approach to their experiment and found: "In contrast to our first study . . . we could find absolutely no specific response to gluten."[19] But once the cat was out of the bag, there was no putting it back in.

One of the most outlandish claims is that gluten intolerance is on the rise because we've genetically modified wheat (creating "GMO wheat"). However, other than the miraculous modifications achieved in nature millennia ago, not a single acre of genetically modified wheat is grown for commercial purposes anywhere in the world. There are a variety of reasons for this, including some unfounded food safety concerns held by some of our trading partners. Carver himself has no problem with genetically modified wheat, but he is sympathetic to some of the concerns that prohibited the initial attempt by Monsanto to introduce herbicide resistance to wheat back in the mid-2000s. Herbicide resistance per se isn't the problem. After all, Carver helped create a nongenetically engineered variety of wheat for Oklahoma that is resistant to a particular herbicide. Carver said this development, along with the introduction of herbicide resistant canola, has finally given farmers a way to effectively combat weed problems that sap wheat yields.

What worries Carver is the privatization of the existing germplasm and the consolidation of all that genetic material in the hands of only a few large companies. As it stands now, Carver is able to collaborate with public wheat-breeding programs in Texas, Kansas, Colorado, and throughout the world to share and exchange genetic material. Genetics is important but so too is an understanding of how a particular set of genes interacts in an environment like Oklahoma's. Carver remarked, "Genetics load[s] the gun but the environment pulls the trigger." Carver worries about whether, when creating new varieties, one large seed company would consider all the particular environmental factors that Oklahoma farmers face but farmers elsewhere do not.

None of this is to say that Carver is against intellectual property protection for plant genetics any more than he is against the protection of creative works in music and art. This protection sustains further creativity and innovation. Opponents of genetic engineering often complain about the ability of private companies to own genes or seeds. Even aside from the issue of whether companies would invest the millions of dollars required to research new developments if they couldn't be assured of a return, these concerns often miss the realities on the ground. Until about 2005, when Oklahoma State University released a new variety—in the form of "foundation seed"—anybody could drive up to the university warehouse and pick it up. The problem is that some folks who picked up the foundation seed would take it home, grow it, and then handle, sell, and redistribute it in ways that did not protect its genetic purity. These third parties were profiting from Carver's decade-long efforts to develop a new variety, often were not thorough in their methods, and would end up contaminating the new variety and selling less productive seeds.

The nonprofit corporation Oklahoma Genetics Inc. (OGI) was created to help protect the intellectual property and fund ongoing wheat genetics research. Individual farmers and companies pay to become members of OGI, which gives them access to the foundation seed. Carver said this helps make sure the seed winds up in the best seed producers' hands. These producers preserve the purity and performance of the new varieties. This set-up also helps fund Carver's research program, allowing him to pay and keep technicians who would otherwise be lured away by higher salaries elsewhere.

Although no GMO wheat is being produced commercially, scientists are studying it. In fact, Carver took me to a plot smaller than my driveway and said, "This is GMO wheat." It looked exactly the same as all the other wheat growing around it, so I asked what was different about it. So he told me the story.

Over the years Carver's traditional breeding program has identified a particular gene in wheat, something he called Lr34, that seems to provide protection against a variety of diseases, like the stripe rust he showed me earlier. The problem is that Lr34 doesn't always work because the gene is not perfectly processed inside the plant. The passing along of genetic information, called the central dogma of biology, is so inefficient for this particular gene that only about 30 percent of the time does a plant with the gene actually produce a protein that can "vaccinate" the plant. In other words, in the process of converting DNA into the protective protein, some translation mistakes occur along the way. One of Carver's colleagues, Liuling Yan, studied the sequence of this gene (which contains sixteen thousand base pairs of A, G, C, and T) and figured out the exact gene sequence that the plant uses when the translation works right. Rather than looking for a variety of seed with this more efficient version of Lr34 and working a dozen years from kindergarten to high school to breed it into a commercial variety, Yan and Carver simply inserted the better, more efficient Lr34 sequence into an existing commercial variety. It took only one year to grow the seniors I saw waving in the wind.

In contrast to transgenic GMOs, which are created by moving genes from one species to another, this was an intragenic (or sometimes called cisgenic) variety created by moving genes within a species. Although this outcome might have been created (much more slowly and less efficiently) through traditional plant breeding, regulatory agencies classify it as a "GMO" because of the technique that Yan used to move the new genes into place. The time and cost it would take to gain regulatory approval for the GMO variety may well be prohibitive, but the scientific tool to help address a major problem facing wheat farmers is there for the using.

Carver is also excited about an altogether different use of GMOs in wheat breeding. One factor that makes breeding wheat different from, say, breeding corn is that wheat is typically self-pollinated. Creating cross-breeds or hybrids in wheat is an arduous task that often entails a lot of manual labor to keep the pollen from one plant from reaching its own flower. Pollen from one plant has to be manually transferred to another plant to make a cross. Whereas most of the corn grown commercially in the United States is a first- or second-generation hybrid produced from crossing two purebred lines, almost none of the wheat commercially grown is a first-generation hybrid because of the challenges in creating a large number of first-generation crosses that can be sold as seed. This means wheat doesn't benefit entirely from the "hybrid vigor" that boosts hardiness and yields when two purebred lines are crossed.

One solution to this conundrum is to use genetic modification to control pollination and fertilization of the parents but to do it in a manner that results in hybrid offspring that are not GMOs. In particular, a genetic modification in one of the parents can render the male portion of the plant sterile, which keeps the plant from self-pollinating. In conventional breeding with plants like corn, where the male and female parts are separated on the plant, hybrids are typically created by manually sterilizing the plant by removing one of the gendered parts (in corn, the process is called detasseling). By turning off the male part of a wheat plant, self-fertilization is stopped, and the possibilities for large-scale crosses begin.[20] This process would allow breeders to produce a large number of first-generation hybrid seed available for commercial use. Although the technology is not in use today to create commercial hybrid varieties, it holds the potential to continue to improve genetic gains.

Growth in yields is a key aspect of sustainability. Yet many environmentalists contend that sustainability means only what nature will allow. The soil is productive until we disturb it and then it declines. In this view a biological carrying capacity limits how many people our planet can support; thus conservation and sustainability will have to mean fewer people. For adherents of this perspective, the solution isn't to increase growth but to limit the number of humans who need to eat. Yet, as the Magruder plots and Raun and Carver's research have shown, the productivity of the soil can be continually increased through human ingenuity, which is perhaps the most important renewable resource. And it has delivered. Data from the U.S. Department of Agriculture show that, for all types of wheat in the United States, yields were about thirteen bushels per acre in the 1930s, jumping to about twenty-five bushels per acre in the 1960s. Since 2005 wheat yields in the United States have averaged more than forty-five bushels per acre.[21] So we are producing more than twice as much food on less land, thanks to better fertilizers, better genetics, and better cultivation practices.

If I had to be honest, in the past I've been a bit curmudgeonly. I often bristled when the word *sustainability* was mentioned. It seemed to be one of those feel-good, vacuous buzzwords thrown around to support whatever cause the wielder of the word wants to champion. Yet it is hard to argue with the general concept of sustainability. Who doesn't want to stay productive over time and have enough resources to enjoy the future?

I've come to realize my problem is not with the concept of sustainability per se but rather with the way many people propose to achieve it. In food and agriculture sustainability has come to be interpreted as synonymous with organic, natural, and local. This perspective posits that the way we endure and sustain our production over time is to have a smaller population, spend more time working the land, spend more money on food, and learn to like to eat different kinds of foods. Maybe that kind of future sounds good to some folks, but if that is the kind of future that will be sustained, count me out.

Our ancestors, at least as a species, could have carried on quite sustainably for a long time, but their sustainable life is not one I'd choose to be born into. The all-natural future is not the kind of future in which I want to live, and I think that is why I've been bothered by the word *sustainability*. The missing ingredient in sustainable thinking is the role of scientific and technological advancement. Sustainable doesn't have to mean stagnant. Rather, any future worth fighting for is one that is dynamic, innovative, and exciting, one in which there will be many other humans with bountiful opportunities to eat and work as their hearts desire.

We don't have to take a step back to sustain our current living and eating standards. We can continue to enjoy the wonderful abundance of food and even improve our living standards. But optimism alone won't cut it. We actually have to be willing to conduct new research, experiment, and adopt new food and agricultural technologies. As history has shown, technological advancement in food and agriculture has been at the root of our gains in prosperity. That's why I now am in favor of sustainability. Because, as I see it, sustainability and using agricultural technology are one and the same.

9

Waste Not, Want Not

Chili is one of my favorite dishes to prepare. It permits a wide range of experimentation, and it's hard to screw up. Plus, you can make it ahead of time and let it simmer until the kids get home or the guests arrive. But what to do with all that leftover chili? I typically cannot bear to throw it out, so it goes into an old Cool Whip container in the refrigerator, where it sits for a week. Then two. Sometimes three. It's then that I start having flashbacks of swabs and plate counts from microbiology class that overpower my guilt, and the chili goes into the trash can.

My ninety-three-year-old grandmother, still affected by her Depression-era upbringing, is far more fanatical. We dare not attempt to dispose of tin foil in her presence. Underneath her sink rests a large ball of tin foil that the cabinet can scarcely contain. A few times a year we have to distract her so we can clear the fridge of once-tasty treats. And opening my mother-in-law's fridge risks setting off an avalanche of old plastic-wrapped cheese and foil-lined missiles.

No one likes the idea of food waste. Yet we're apparently quite good at generating it. One U.S. Department of Agriculture report estimates that more than 30 percent of the nation's food supply—133 billion pounds each year—goes uneaten.[1] We could make a serious dent in the food security problems that plague many parts of the world if we could figure out how to make better use of the food that's already produced. In the developing world, food waste often results from inadequate infrastructure. Grain storage facilities (to the extent they exist) don't keep out the pests or the weather. Trucks are unable to navigate the roads (to the extent they exist) quickly enough to move crops from where they're grown to the people who want to eat the crops before they've rotted.

Here at home, food is often so inexpensive that, economically, it is hard to justify holding on to it for long. Some foods, like fresh fruits and vegetables, have relatively high rates of waste because they become overly ripe so quickly. We already use many technologies that reduce food waste by extending taste and shelf life—sometimes by using natural preservatives (like salt or vinegar) or unnatural ones like refrigeration, canning, or sodium benzoate. Yet we have a sinking feeling that we throw out more than we should.

Where there is waste, there is opportunity.

Eldon Roth was on the lookout for opportunities. He grew up poor in the 1940s and 1950s in rural South Dakota. His earliest memories are of a home that didn't even have a refrigerator.

When his parents bought their first refrigerator, it wasn't the electric sort that we are now accustomed to seeing. It ran on kerosene. Roth recalls being awestruck by a flickering blue flame that could somehow make the contents of the refrigerator cold. By heating a refrigerant with a low boiling point, the flame extracted heat from the refrigerator compartment, lowering the temperature to a point where spoilage could be slowed or stopped. Of course, Roth didn't know that at the time.

Roth never got a college degree. But he had a curious mind and a tinkering spirit. Roth landed a variety of blue-collar jobs in the food industry in California in the 1960s. One of those jobs was in a cold storage plant that froze beef cuts soon after slaughter.

The trouble with refrigeration, Roth noticed, was that it, too, can be wasteful. The process takes time and often requires cooling a large warehouse. Ultimately, the faster meat can be cooled and the longer it can be held at a low temperature, the safer it will be and the longer it will last. If you've ever watched the scene in the 1980s classic *A Christmas Story* in which Ralphie's poor friend, Flick, is dared to lick a flagpole in winter, you know that cold metal freezes more quickly than air. Roth had the idea of running meat through chilled metal rollers to more quickly bring its temperature down to a point where spoilage was unlikely. His new flash freezing process reduced freezing time from a matter of days to just a couple minutes.[2]

In 1981 Roth founded Beef Products, Inc. (BPI) and during the next thirty years opened four food-processing plants across the Midwest that made use of his roller freezing technology.

Looking to expand, Roth eyed the massive hamburger business and saw an opportunity.

Before it heads to the packing house, a typical live steer weighs about 1,300 pounds. After the carcass is dressed and its head, hide, and organs are removed, about 850 pounds of useable meat and bone are left. When we think beef, we often think steak. But if ranchers want to sell steak, they've got to do something with all the rest. Only about 20 percent of the dressed carcass is comprised of steaks. The other 80 percent are cuts like roasts, ribs, and flanks. All in all, almost half the carcass ultimately winds up as ground beef.

American Indians are often held up as exemplars for their respectful and careful use of every part of the buffalo; they let nothing go to waste. Meatpacking houses face strong economic incentives to do the same. In the late 1800s, Gustavus Swift, for example, would personally trudge out in his dark suit and top hat to inspect the sewer pipes that flowed out of his Chicago packing plants into nearby rivers, not because he was an environmentalist but because any trace of fat or hair meant waste and lost revenue.[3] Swift fought waste in other ways. He revolutionized the meatpacking industry with the use of refrigerated railcars. Before then, urban meat eaters had to deal with the squalor and hassle that came with livestock herded along city streets to nearby packing houses.[4] Refrigeration allowed the packing plants to move far outside town and still provide safe, fresh meat to shoppers. And because it helped prevent waste, refrigeration yielded a more plentiful food supply. Despite the incentives to reduce waste and increase efficiencies, it is often difficult to find productive economic uses for every morsel.

Such is the case with beef trimmings. When dividing the carcass into various primal and subprimal cuts, pieces often remain that are difficult to sell on their own. About 15 to 20 percent of the meat from a beef carcass can end up as trimmings, which are often used to make ground beef. But some trimmings are too fatty to be of much use in the process, particularly as consumers have come to demand leaner ground beef in recent decades. It is possible to take out a knife and separate the strips of lean from the fat, but it is a laborious process and one that doesn't pay, given the cost of labor. The result was that lots of perfectly edible beef was rendered useless and not eaten by humans.

The challenge was to find an economical method to pull out the nuggets of lean from the fat. Roth turned to the same insight that made his parents' kerosene-powered refrigerator work: different substances change from solid to liquid to steam at different temperatures.[5] Fat becomes liquid at a temperature lower than that needed to cook protein. By warming the fatty beef trimmings, Roth found that the high-value edible protein could be pulled away from the less valuable fat. Once the fat is softened, BPI uses a centrifuge to separate the two substances. It then sells the fat on the tallow market, and the protein—the same protein that was in any other roasts or beef trimmings—can be used in, well, anything that is made of beef protein, like hamburger.

The problem Roth had to solve was that although the protein remained uncooked, its temperature had risen to a point that made the beef susceptible to spoilage. Good thing Roth had his roller freezing technology. After being passed over the chilled metal rollers, the product becomes known as "lean finely textured beef."

BPI ships the lean finely textured beef to other processors, food manufacturers, and restaurants where it is typically mixed with other cuts and beef trimmings (often at a ratio of one pound of lean finely textured ground beef to seven to nine pounds of beef trimmings) to make hamburgers and lean ground beef. Oklahoma State University research shows that consumers like the taste of burgers that contain lean finely textured beef just as well as they like burgers that don't have it.[6] Because BPI's product is 95 percent fat free, it is used to help meet the demand for 90 percent and 95 percent lean ground beef, reducing the price of the 90 percent lean ground beef sold in grocery stores by about 4 percent.[7]

Chances are that if you've eaten a hamburger in the last twenty years, you've enjoyed the fruits of Roth's labor.

Perhaps more important than the price reductions Roth's process has helped deliver is the reduction in waste. The renowned animal welfare expert Temple Grandin told an interviewer this about Roth's process:

> Let's say we stopped using [lean finely textured beef]. That would be the equivalent, at this great, big plant, that we just take one entire truckload of their cattle, we just take them to the dump and shoot them and throw them in the dump. We'd take a truckload each day from each big plant and just throw them in the garbage. That would be about 42 head of steers,

and we're going to just throw them away. That's really unethi-
cal, that's food waste.[8]

BPI itself has argued that if we stopped eating lean finely textured beef, it would be equivalent to "throwing away 5,700 cattle
a day."[9]

I visited Roth on a cold fall morning at his plant in Dakota
Dunes, South Dakota. He was calm and relaxed and looked more
like a grandfather than trend-setting entrepreneur. Roth's son-
in-law, Craig Letch, who heads up safety and quality for BPI,
described the intricate workings of the plant, where every pound
of pressure, all temperatures, and every product line can be moni-
tored from a computer screen in the conference room. Between
discussions of air flow and centrifugal force, Roth would occasionally chime in with a warm story or provide some context for
why a particular part or process had to be developed.

The plant itself is a technological marvel and much of the
equipment was made in house, designed by Roth and his team.
The freeze rollers were massive, shimmering stainless steel con-
structions. The slow-turning rollers were several stories high,
with just enough frost on the edges to let you know they'd freeze
anything that touched them. The protein is automatically moved
from the centrifuge to the rollers, where it quickly freezes before
being packaged. In a world in which safety and cleanliness are
paramount, no human hand ever need touch the food from the
time it enters till it leaves the plant.

Safety. Cleanliness. Those were words that, as you would expect, BPI employees used frequently. But all the talk really wasn't

necessary. From looking around, it was obvious that these were core values for the company. The air is so heavily pressurized (to prevent contaminants from entering), that you have to literally pull yourself into the plant. I nearly fell down as the pressure threw me out the door when I left. The air is literally washed—in huge, custom-made stainless steel machines that pull the air through disinfected water—before being recirculated.

After I heard Roth talk, the emphasis on safety was clear. He feels responsible for the food that leaves his plant—for the safety of the people who eat burgers that contain his product. He became almost teary as he talked about the parents he'd met after the Jack in the Box tragedy. In the early 1990s more than six hundred people became ill and four children died after they ingested a deadly strain of *E. coli* from a Jack in the Box burger. Roth had nothing to do with the tainted meat used by Jack in the Box, but what happened made him understand what awesome responsibilities he had.

Roth's focus on safety has paid off. To his knowledge BPI has never sold an ounce of product that has made someone ill. In 2007 the International Association for Food Protection awarded BPI its highest honor in recognition for its commitment to food safety. In 2008 *The Washington Post* profiled the company and highlighted its focus on safety. Roth also received the Beef Industry Vision Award in 2008 from the National Cattlemen's Foundation. In early 2012 Roth was inducted into the Nebraska Business Hall of Fame.[10]

Soon after Roth accepted that last award, his life changed forever.

I suspect that many readers are unfamiliar with the term *lean finely textured beef* but are probably more familiar with the more insidious one that became affixed to Roth's product: *pink slime*.

It is catchy but inaccurate. The product is not slimy. And it is pink only when frozen (as is the case for all beef). Google "pink slime" and you are almost certain to find a picture of a yogurt-like substance that is decidedly *not* Roth's lean finely textured beef. I've yet to find a food industry expert who knows what's in the picture, yet it is the web image most viewed in connection with the term.

It's hard to know exactly how the rumor-filled, social-and-conventional media firestorm started. The company, proud of its humble origins, technological innovation, and safety record was happy to open its doors to visitors and the media. They were even featured in the widely acclaimed 2008 film *Food, Inc.* After giving the documentarians access to their facilities, the film showed the company in dim lighting with ominous background music. Ironically for a food film that advocates transparency, the filmmakers presented a slanted view of the company without appropriate context or nuance. Despite the documentary's popularity, there was little sign of public uproar—perhaps because the term pink slime had yet to be introduced to the public. The company received more attention when Michael Moss, a reporter for *The New York Times,* mentioned BPI in a story about a beef safety recall (a recall that ultimately had nothing to do with BPI's product). Moss, who would go on to write the 2013 best-seller *Salt, Sugar, Fat: How the Food Giants Hooked Us,* followed up with a *New York Times* exposé specifically focused on BPI's product.

The issue that seemed to raise eyebrows was a procedure sometimes used by BPI: spraying a light mist of ammonium hydroxide on the lean finely textured beef. Ammonia is a substance that occurs naturally in many foods, and it is an ingredient that the Food and Drug Administration lists in the category of "generally recognized as safe." BPI used ammonium hydroxide (ammonia mixed with water) because it lowered the pH of the beef and helped prevent bacterial contamination. As I learned on my visit, the ammonia mist is not always applied, and its application depends on the customer's requests and order specifications. It turns out that there is more ammonia in the cheese and in the bun of a cheeseburger than there is in a hamburger patty made with lean finely textured beef.[11] But to a general public that is more fearful of sodium chloride than salt, the story seemed to suggest a *Jungle*-like narrative that fed the public's distrust of processed foods and their makers.[12]

It didn't help when, in April 2011, the celebrity chef Jamie Oliver invited a group of schoolchildren and their parents to the premier of the second season of his ABC show, *Food Revolution*, to see his take on lean finely textured beef. In his characterization of the process, Oliver placed cuts of beef he called "dog food" in a washing machine. He then drowned the beef with a jug of ammonia that bore a prominent label displaying a skull and crossbones. The audience responded predictably to his question: "Do you want it fed to your children?" Without any explicit reference to Roth or food waste, Oliver said that "everything about this process, to me, is about no respect for food or people or children."[13]

The larger food-consuming public essentially ignored Oliver's show. But then a lawyer-writer-blogger-mother started a petition to remove lean finely textured beef from school lunches, and *ABC World News with Diane Sawyer* ran a series of reports on BPI in March 2012 that resulted in a public outcry. Despite the earlier stories by *The New York Times* and Jamie Oliver, ABC touted the story as coming from a whistle-blower. BPI's major buyers—many of whom were named in ABC's reports—stopped ordering.

BPI eventually closed three of its plants and had to lay off more than six hundred employees. Finely textured beef went from being an ingredient in almost three-quarters of the nation's hamburgers to nearly none. Nancy Donley, one of the mothers of a child who died in the Jack in the Box *E. coli* outbreak, wrote an op-ed defending BPI and its reputation for food safety. But it was too little too late. *Bloomberg Business Week* reported that Roth was heartbroken as the company he had struggled to build for three decades crumbled around him.[14]

It's a bit hard to know what to make of all that transpired. To be sure, much of what was said about BPI was sensationalized. BPI didn't use organ meats or bones or hoofs or hides or "dog food." The company used slightly fattier versions of the same beef cuts that usually become roasts or ground beef. In fact, the day I visited BPI's South Dakota plant, which is adjacent to a Tyson packing facility, I was amazed at the beef entering BPI's facility. The meat traveled on a conveyer belt in a tunnel that connects BPI and Tyson. A steer or heifer enters one end of the Tyson facility, and a few hours later beef trimmings emerge at BPI without ever seeing the light of day. The trimmings consist of some small cuts of beef, but there are also huge hunks of meat that looked almost identical to the briskets that I love to barbeque for get-togethers with friends and family. Lean finely textured beef is beef. That's all. I suppose that's why the company created a website called beefisbeef.com. No bone goes into the process. Big beef hunks go in one end and out the other end come three products: tallow, cartilage (which is the only waste), and lean finely textured beef.

I've visited a lot of food plants, and BPI's was one of the most technologically advanced, safety-conscious plants I've seen. That a company that proactively invested millions in food safety measures found itself embroiled in controversy involving perceived (but unfounded) safety concerns is deeply ironic. What tarnished BPI's reputation was no actual sickness or recall or outbreak; it was a series of TV shows and news stories.

But, given the information that consumers received, it is hard to fault them for their reaction. After all, best-selling authors and journalists have primed the public's distrust of Big Food. In an era when processed food has come to be seen as almost evil, "pink slime" struck a chord with consumers. Perhaps BPI should have required labeling of the beef that contained its products. Surely some of the public outcry arose from a feeling of having been deceived and of having no control over what is in our food. But from BPI's perspective, what's to label? "This product of ground-up beef parts contains more ground-up beef parts"? More fundamentally, BPI didn't sell directly to consumers. Rather, the company sold to other processors, who sold to restaurants and grocery store chains. BPI was hardly in a position to force others to label products that contained lean finely textured beef.

So where does that leave us? Many shoppers, although I am not among them, no doubt want to avoid lean finely textured beef and are willing to pay a premium to purchase lean ground beef that does not contain it. There's no harm in that.

But if we are really concerned about food waste, we probably need to change some of our narratives. We shouldn't say we want companies to recycle and reuse and then turn around and vilify them for doing so.

The comedian Jon Stewart, who was more than willing to jump on the Big-Food-is-bad bandwagon, remarked that pink slime should instead be called "ammonia-soaked centrifuge-separated by-product paste."[15] He was working off a popular narrative. He could have instead featured the harm to a family-owned business that was innovating to make food safer and more affordable by preventing food waste. But that's not very funny.

10

Food Bug Zappers

When I was living in Paris a few years ago, I would often walk past the École Normale Supérieure, a leading French university where Louis Pasteur once worked. Once I stopped to look at a plaque celebrating Pasteur outside a nearby school. When I turned around, I noticed a natural food store that sold organics, herbal remedies, and artisanal whole grains. I was startled by the irony of seeing a large sign in the store window touting raw milk. When Pasteur discovered that heating milk (and wine) to high temperatures could kill bacteria in a process that came to be known as pasteurization, he probably couldn't imagine a day when it would be necessary to seek out obscure specialty shops to find milk that did *not* use the process he created.[1]

Pasteur's work has had immense consequence. The Centers for Disease Control and Prevention (CDC) estimates that milk pasteurization leads to a 150-fold reduction in the number of outbreaks of food-borne illness.[2] Before vaccines and other treatments for tuberculosis became available after World War II, an eradication program that included milk pasteurization—the disease spread to humans from infected cows—was estimated to have prevented at least twenty-five thousand deaths annually in the United States.[3] Some people like to drink raw milk because they think it tastes better or because they believe it has health benefits, but no one can deny that Pasteur's process has saved lives.

Pasteur's work helped solidify the germ theory of disease— the idea that some diseases are caused by microorganisms. Previously, people thought illness, along with food and water contamination, was caused by spontaneous generation, bad air, or by gods and evil spirts. Even today we tend to forget or don't realize that invisible, illness-causing critters are traveling in the air, on our bodies, and in our food.

A common view seems to be that food safety problems are a modern phenomenon brought about by the development of the industrial food system. But killer food microbes have been with us since the beginning, of course. Almost all major religions have various food taboos that might be explained, at least in part, by efforts to avoid food poisoning. Pork, an unclean food for Jews and Muslims alike, was a common carrier of the roundworm Trichinella until recent years when hogs were moved to more sanitary environments with diets that were more carefully controlled. We know bacterial contamination and food spoilage was a big problem throughout human history because our ancestors

devised many methods to prevent it. The ancients would introduce chemicals, through salting or smoking, or remove enough
water through drying to inhibit microbial growth. Fermentation
would result in enough acid or alcohol to keep the microbes at
bay. Salted fish, dried figs, beef jerky, cured ham, and wine all
are results of attempts to extend the shelf life of food and make
it safe to eat.

Using preservatives to prevent microbial contamination is as
old as civilization. In *Food in History* Reay Tannahill says this
about the wine drunk by ancient Greeks:

> There seems little doubt that the wine had a characteristic
> tang that might not find favour today. It was fermented in
> vats smeared inside and out with resin, and the goatskins
> or pigskins into which it was subsequently filtered no doubt
> made their own contribution to both flavor and aroma. Since
> fermentation was not a scientifically controlled process, the
> unadulterated wine did not keep well, and by early Classi-
> cal times most regions had developed their own additives
> to rectify this. One formula involved a brew of herbs and
> spices that had been mixed with condensed sea water and
> matured for some years; another used liquid resin blended
> with vine ash and added to the grapes before fermentation.
> Wine was often matured in the loft where wood was seasoned
> and smoked.[4]

Adding preservatives and antimicrobial agents like salt, herbs,
spices, smoke, resin, and ash to wine hardly seems natural, even

though some of the ancient world's greatest philosophers probably drank the stuff.

The modern-day quest for naturalness in food sometimes
runs directly at odds with food safety. Part of the problem reflects a simple lack of knowledge and what might be called
chemophobia—a fear of chemical-sounding words. A couple
years ago I polled more than a thousand U.S. consumers, gave
them a list of eleven ingredients, and asked whether they thought
a food with each ingredient would be considered natural.[5] More
than 70 percent of respondents thought a food with added cane
sugar is natural, and about 66 percent thought the same of a food
with added salt. When I asked about added sodium chloride,
only 32 percent thought such a food would be natural. Yet sodium chloride is simply the chemical name for salt. I halved the
number of people who thought a food was natural just because I
had used salt's chemical name instead of its colloquial one!

The bigger problem, however, is what happens to the safety
of food when seemingly unnatural ingredients are not used.
Keeping food safe without using chemical additives is a big challenge for food manufacturers and retailers. Consumers are increasingly demanding fresher, more natural, "clean" food. Yet,
as one food safety expert told me, "It's a tremendous strain on
the food-producing industry. If you take away growth inhibitors,
what do you do?" One executive of a large food retailer remarked,
"As consumers are asking for fresh and more natural food, we
have to take out ingredients and preservatives, which makes food
less safe."[6] Fresh foods might have taste advantages, but they also
tend to have shorter shelf lives, increasing the likelihood of earlier

spoilage and food waste. Moreover, research and development costs involved in reformulating preservatives to increase the perception of naturalness are passed on to the consumer in the form of higher food prices, even when the preservatives' underlying chemical properties have not changed.

For example, early in the twentieth century, scientists learned that the active ingredient in the saltpeter that had been used to cure many meats for centuries was nitrite, and food processors began to replace the saltpeter with smaller quantities of sodium nitrite. When modern consumers began to express concern about the use of "unnatural" sodium nitrite in products like hot dogs, food processors were unsure what to do. Making certain kinds of cured sausages and hot dogs without using some kind of preservative is nearly impossible. Ultimately, many food manufacturers responded by using natural extracts from vegetables like celery that are high in nitrates (which the body converts to nitrite). This allows the companies to use the label "no added nitrites," but the resulting cured meats are not free of nitrites. Switching to more natural ingredients might ease the consumers' conscience and allow retailers to sell more product, but the reformulation has virtually no impact on the ultimate chemical makeup of the food. As it turns out, the vast majority of nitrates (and thus nitrites) that enter our bodies are not from cured meats. We get four times the amount of nitrites all naturally from vegetables like arugula, beets, lettuce, celery, and spinach and from herbs like basil and coriander.[7]

Regardless of how you regard nitrites or the benefits of more "clean" food, attempts to sell fresher, more natural food will, clearly and perhaps ironically, require human intervention to ensure that the food does not spoil and is safe to eat. This isn't just a reformulation problem; making sure the foods we eat every day are safe and wholesome is a serious and important challenge. The CDC estimates that more than 15 percent of Americans suffer from food-borne diseases each year, resulting in 128,000 hospitalizations and 3,000 deaths.[8] Food-borne illnesses result in death, hospital bills, and lost time at work, which is estimated to cost Americans about $14 billion each year, an amount that escalates to $77.7 billion per year if the monetary value includes the hassle and pain and suffering from being sick.[9] Most experts think these estimates underestimate the toll of unsafe food because most food-borne illnesses are never reported. That little tummy ache, mild stomach flu, or bout of diarrhea probably isn't a mere accident but is often the result of eating contaminated food.

Fortunately, a new generation of Louis Pasteurs is at work. As I'll discuss, they're finding new ways to zap bugs, creating devices to rapidly detect contaminates, using information technology to check proper food handling, and much more. But why would food companies do this? Isn't cutting corners a corporate technique for boosting the bottom line?

As it turns out, the story is far more complex. To get a glimpse of the world of food safety at a major food company, I called Kevin Myers, the senior vice president of research and development for Hormel Foods. After stints with Oscar Mayer and Sara Lee, Myers now leads teams responsible for developing new products at Hormel Foods. About one half of his team focuses on food safety validation of new and existing Hormel Foods Corporation products.

I asked Myers about the role of food safety at a company like Hormel Foods. He said, "Large food companies have a brand name. This is what allows them to have higher earnings over generics. Protecting that brand is utmost." He pointed out that Hormel Foods has more than thirty brands that are either number one or two in their category in terms of sales. Hormel pulled in more than $9 billion in revenue in 2014 from sales of brands like SKIPPY® (peanut butter), MUSCLE MILK® (a line of protein nutrition products), SPAM® (meat products), and HORMEL® Pepperoni. They have also recently acquired Applegate Farms (natural and organic meats). The reputation of brand names might allow firms to make a bit more money, but it also exposes them to large potential losses in the event of a product recall or food safety problem. Reputation is a two-way street, and a once-solid name can quickly work against you if it becomes tied to bad news. Research shows that meat recalls by publicly traded companies typically result in a 1.5 percent to 3 percent loss in shareholder wealth.[10] For branded products like hot dogs, a food safety recall tends to reduce sales by more than 20 percent, and the negative effects persist for more than four months.[11] But for Myers, who grew up on a small farm in Iowa, it's about more than just the bottom line. "These are the same brands we buy for our families every day," he said. "We don't want to take any shortcuts."

If family bonds, neighborliness, and financial incentives aren't enough, the recent criminal convictions of the brothers Michael and Stewart Parnell surely got the attention of food industry workers. Following the deaths of nine and the illnesses of more than seven hundred people in 2008 and 2009, investigators linked the outbreak to contaminated peanut butter from a Georgia food-processing company owned by Stewart Parnell (Michael Parnell was the company's food broker). According to one news source, "Never before had a jury heard a criminal case in which a corporate chief faced federal felony charges for knowingly shipping out food containing *Salmonella*."[12] In the wake of the outbreak, not only did the company file for bankruptcy, the two brothers were sentenced to more than twenty years in jail. A quality control manager at the food plant will also spend five years in prison.[13]

The Parnell brothers, who knowingly shipped contaminated food, were a couple bad apples. Many others in the food industry are not only seeking to put out safe food but are spending millions to further reduce the risk of food-borne illness. Myers told me about one of the innovations that Hormel Foods helped pioneer. Back in the 1990s, many food companies began to adopt a food safety system known as HACCP (which stands for hazard analysis of critical control points) that became mandatory for meat companies in the latter part of that decade. HACCP focuses on the prevention of pathogens (microbes that cause illness) by identifying and monitoring so-called critical control points (CCPs). A common CCP for many food companies is the cooking stage, and companies follow a process of taking the temperatures of each batch to ensure that the product is raised to a temperature high enough and for a long enough period to effectively kill pathogens.

Hormel Foods sells a number of meat products, like prosciutto, which did not have a well-defined CCP when initial HAACP assessments were conducted. These meat products are traditionally made safe by drying and salting, and while the company could check for water activity, pH, or appropriate application of curing ingredients, Hormel Foods researched additional methods to have a definitive CCP kill step. Myers said they looked at several options, including irradiation—exposing food to radiation, which kills bacteria. While the technology is perfectly safe and has been around for decades (the U.S. Army has used it since the 1950s), consumers have never warmed to the idea. Today irradiation is used to control contamination in spices, but regulatory hurdles have hampered its application to other food products (Myers said the FDA still hasn't approved irradiation for use in multiple-ingredient products). Also, mandatory labeling rules force manufactures to display the Radura on treated products—and the symbol scares away many would-be buyers.

Myers said Hormel Foods also tested, and ultimately implemented, high-pressure processing, a technology first proposed in the late 1800s but that had not been used to a great extent in commercial applications until recently. The problem that confronted Myers and his team is that bacteria are everywhere—on animals, in the air, on equipment, and on workers' and consumers' hands. Any cut, sliced, or ground meat product could have bacteria not only on its surface but also on the inside of the product. An effective preventative technology needs to kill microorganisms both outside and inside the product. Cooking would do the trick, but cooking changes the taste and texture of products like prosciutto, and even for products that are cooked, bacterial contamination can occur after cooking if the product is handled (unless it is canned or jarred).

High-pressure processing (sometimes also called pascalization after the seventeenth-century scientist Blaise Pascal, who studied pressure) allowed Hormel Foods to sanitize both the meat and the package it comes in. The process is particularly well suited for ready-to-eat foods because it takes place after the product is packaged and eliminates potential contamination that could occur after cooking and before packaging.

Myers said the process works by placing the packaged food in a chamber and submitting it to extreme levels of pressure. Have you ever jumped off the diving board at the deep end, only to have your ears hurt as you approached the bottom of the pool? That pain is caused by the pressure exerted on your eardrums by the water above you in the pool. At a depth of about ten feet, your ears are feeling about 4.3 pounds per square inch (psi) of pressure. If you could somehow swim to the deepest point in the ocean (about thirty-six thousand feet down), you'd feel more than 15,600 psi. Well, you wouldn't actually feel anything because your body would be crushed well before you reached that depth. According to Myers, Hormel's high-pressure processing system applies 87,000 psi to food products. That is five and a half times more pressure than would be felt at the deepest depth of the ocean.

Foods subjected to high-pressure processing enter the chamber prepackaged (often in plastic vacuum packaging). Then cold water is pushed into the chamber, and pumps are turned on until the pressure is ratcheted up to the desired level. Within a few minutes proteins (including bacterial DNA) held under such high levels of pressure lose their structure (they're denatured), and cell walls of vegetative bacteria are obliterated. You might expect that the package would collapse under such high pressure, but because pressure is applied equally on all sides of the product, the packaging remains intact. Moreover, the pressure is applied not just on the outside of a food product but equally throughout, which means that it kills the bacteria on the interior of the food as well as on the surface.

Because the process wipes out pathogens without heating, it is sometimes called cold pasteurization. Still, Myers noted that because the process denatures proteins, some fresh products can take on a cooked appearance after high-pressure processing. For example, it would be possible to "cook" an egg—hardening the yolk and sanitizing the product—even when the temperature is near freezing, he said. Nonetheless, researchers at Hormel found that the process worked well for cured meats, and taste panels decided that the process did not affect the sensory aspects of the product.

Other companies have since adopted and adapted the process (using different temperatures and degrees of pressure), and it is used today on products like guacamole and yogurt. In fact, third-party contractors now specialize in high-pressure processing. A food manufacturer can ship its products to the contractor and pay to have its foods treated before they're shipped to grocery stores, restaurants, and the final consumer.

One big area in which it is used today is fruit juices. Because of perceived taste and health benefits, a number of companies have attempted to market unpasteurized fruit juice. However, in 1996 a batch of the company Odwalla's unpasteurized apple juice containing *E. coli* O157:H7 killed one girl and sickened at least sixty-six others. The company was held criminally liable and paid $1.5 million in fines.[14] The tragedy emphasized the importance of pasteurization, but some consumers don't like the taste of juice that has been subjected to high heat. As a result a number of juice companies have moved toward high-pressure processing in recent years in an effort to control bacteria while retaining fresh flavor.[15] There is a lot of debate about the proper labeling for these juices—natural or fresh?—but there is little doubt that they're safer to drink.

High-pressure processing isn't a panacea. Care must be taken to ensure that the process does not affect the taste or texture of the food. Moreover, the process may not kill every last trace of harmful bacteria or bacterial spores.[16] It is a pasteurization process, not sterilization, so products must still be refrigerated. Still, the process quite clearly increases a product's shelf life. And it does so without introducing food additives or antimicrobial agents that might affect flavor and perceived naturalness. The process also sanitizes both the food and its packaging. As a result the process kills whatever bacteria were on workers' hands or the company's equipment before the product was sealed in its final packaging.

Research on food safety techniques continues on other fronts as well. Restaurants and food manufacturers also want to be able to test food for contamination.

Before 1980, if a food manufacturer wanted to know whether its product was contaminated, the company had to create a microbial culture. In the mid-1990s, when I took college microbiology courses, this is how I learned to test for food contaminates, and many of the food plants I visited around that time used the same method. You would take a sample of food, grind it up, and then, using an industrial-sized Q-tip, spread the puree across some gel in a petri dish. The gel (or agar, as it is often called) contains nutrients to feed any microbes present, as well. To encourage the growth of any bacteria harmful to humans, you would often place the petri dish in a warmer that, not coincidentally, was often set at about 98.6 degrees Fahrenheit. And then you waited. And waited. And waited for the microbes to grow.

After several days you'd take the petri dish out of the warmer and look to see what had grown. I have striking memories of the first microbiology class I took. The instructor asked me and my classmates to place our hands on the agar gel in petri dishes that had been placed on our lab benches. A week later we returned to the lab to find our petri dishes contained gnarly black, yellow, white, and green furry monstrosities growing in the precise shape of our hands. That petri dish was easily the most vivid demonstration of the importance of hand washing I've ever seen.

We didn't attempt to identify the dozens of different types of bacteria, yeast, and molds in our handprints, but a food manufacturer needs to know precisely what's in a food. So, back in the day, typically when a cultured petri dish was removed from the warmer, a trained microbiologist or technician had to visually inspect the dish for different bacteria. In college we spent whole classes learning to identify different types of bacteria like *Sal-monella, E. coli,* and *Listeria.* We would also place the petri dish under a transparent grid and count the number of cells growing in each quadrant so that we could calculate the extent of con-tamination in the food. For many microbes the old adage "the dose makes the poison" is apt. Eat a few cells of some kinds of *Salmonella* and you'll be just fine.[17] Eat a few million, and in a few days you'll be miserable.

Here's the key point about that old technology: it could take up to a week to identify whether a food was contaminated with harmful bacteria. If you're making canned soup or jarred salsa, it might be possible to hold these products in a warehouse before distributing them to grocery stores or restaurants while waiting for the test results. However, holding large inventories, even for products with a long shelf life, is costly. As a result many manu-facturers ship their products to consumers and then issue a recall if a test comes back affirmative for a pathogen. Recalls are costly. And they're not 100 percent effective. Sometimes recalls are is-sued after consumers get the contaminated products home, and some smaller retailers may be unaware that they're stocking con-taminated wares.

If the time lag associated with culture-based microbe-testing technologies was problematic for long-lived canned foods, it was an even bigger problem for foods that need to be eaten soon after they're produced. No one likes the idea of eating weeks-old bread, milk, or ground beef. It was hardly possible to learn whether these fresh products were contaminated before we ate them. In fact, the longer you waited on the test, the greater the chance that spoilage would occur in the meantime—the very thing we want to prevent.

Fortunately, a series of technological developments have dramatically reduced the time it takes to test foods for microbial contamination. These rapid detection methods also are more accurate and are easier to use than the old culture-based testing methods. To learn a bit about these new technologies, I talked to Stan Bailey, the senior director of scientific affairs for bioMerieux, the largest food and industry microbiology diagnostics company in the world.[18]

Much of the food you routinely eat has almost certainly been tested with the technologies made by bioMerieux. The company's founder, Marcel Merieux, was a student of Louis Pasteur. Merieux's original company focused on developing veterinary vaccines in the late 1800s, but today the company has annual revenues of more than $2 billion and focuses mainly on diagnostic solutions in more than 155 countries.

Bailey has a distinctive southern drawl, and when I asked about his background, he told me he had been a marine biology major at the University of Georgia before making his way to the field of environmental health. Initially, he was just looking to finish school and get a job, but he grew fascinated with how microbes grew and interacted with the environment. More important, he said, was that "like most people in microbiology, I felt like I was doing something good by improving public health." In his role first at the U.S. Department of Agriculture and now bioMerieux, he said he's worked with many people in the food business. "People in the food industry are good people" who "want to do the right thing," Even if we buy from companies full of good people, we may still want to verify that their products are safe. And testing ensures that they are using the right techniques to ensure food safety. The more accurate and rapid the test the better. And that's precisely what bioMerieux is working to provide.

When I asked Bailey about the benefits realized from more rapid microbial testing, he noted the improved ability of companies to prevent recalls and ultimately human illness. He also mentioned that more rapid microbial testing is now in high demand because of increasing calls for "natural" food, which has a shorter shelf life. Government regulation has been another motivating factor. Meat producers began facing a zero-tolerance limit for the deadly pathogen *E. coli* 0157:H7 in 1994 and then several other *E. coli* strains in 2012, meaning it is illegal to sell product with even a trace amount of what are now considered adulterants. Ground beef, however, needs to be sold quickly after production. BioMerieux has responded with new technologies and processes that have dramatically reduced the time it takes to detect and identify deadly pathogens. Determinations that took four to seven days in the 1980s now take less than a day. And real-time detection systems that work in ten minutes or less are starting to appear.

One technology that began to appear in the 1990s was the immunoassay. When bioMerieux first offered the technology with tests for *Listeria* and *Salmonella,* the process took forty-eight hours. Over time the company has added tests for additional pathogens, and it continues to perfect and simplify the process so that it now works in about eighteen hours. While newer, faster technologies are available, Bailey said that immunoassays remain a "major workhorse" in the industry. The main advantage of modern immunoassays, in addition to their speedier response times than older technology, is that they are almost all automated, Bailey said. A food processor does not need to hire a molecular biologist or send samples off to a lab. A technician has to touch the sample only twice for a total of thirty seconds. Bailey noted that the success of a technology like immunoassay rests not just in technical precision and accuracy but in its cost effectiveness and the ease of use by the end user. If we want safer food, we need technologies that food company employees can quickly and easily use and that are not frightfully expensive.

A testament to the way technology has led to cost-effective, easy-to-use immunoassays is that you can easily buy one in any major grocery store. In fact, you might even have used one yourself without realizing it. An over-the-counter, pee-on-a-stick pregnancy test is an immunoassay. To explain how they work, I need to step back at bit.

When you're sick, your body fights infection by creating antibodies. Antibodies come in different shapes and sizes, and each type is designed to fight a particular infection. If you breathe in the chicken pox virus, your body will send out chicken pox antibodies to fight the infection. If you've got a case of strep throat, your body will send out antibodies that target the offending streptococcal bacteria. Each antibody knows which foreign invader to attack because it is uniquely made to match and bind with that particular type of intruder. Chicken pox antibodies won't attack the bacteria that cause strep throat.

Scientists have learned to design tests—immunoassays—that make use of this precise matching of antibodies and invading bacteria. An immunoassay for food applications would include antibodies that look to bind with, say, *E. coli* or *Salmonella*. If a match is made, another molecule will produce a signal (often by fluorescing or by changing color) to let the technician know a bacteria has been detected. In the case of pregnancy tests, the stick includes an antibody that looks to bind with a particular hormone that women produce after an egg has been fertilized. If the antibody is able to locate the hormone in question, a signal (often in the form of a little blue line) appears on the stick to tell you that your life is about to change forever.

Bailey noted that developing new diagnostic technologies for food safety is, in many ways, much more challenging for food than it is than for medical applications. When trying to identify whether someone has a particular disease or infection, a doctor is typically dealing with an already sick patient, which means large numbers of the invaders are present for tests to detect. With food, however, pathogens are typically not present in large numbers (if they're there at all). This means food tests have to be much more sensitive than medical tests. Or it means food bacteria need more time to grow to a detectible quantity.

An important distinction in food safety testing is between detection and identification. According to Bailey, detection (that is, are *E. coli* present?) can be performed relatively rapidly, but identification (what kind of *E. coli* are these?) has historically taken longer. The distinction is important because only a few of the more than two thousand types of *Salmonella* will make you sick. Use of immunoassays has brought down the identification times. Other developments, like polymerase chain reaction (PCR) technologies, are more expensive but are quicker and more accurate. PCR works by replicating (or amplifying) the DNA from potential pathogens. Because these pathogens are often present in small numbers, the method provides a practical means of identifying hard-to-find bugs. Like immunoassays, PCR uses specific primers, or probes, that bind with the DNA of pathogens and send signals if a match is made. The whole process can be completed in two hours. Even more expensive and rapid technologies, such as mass spectrometry, are being brought to bear on food safety testing, and these machines can be used to test not only for harmful bacteria but for a wide array of potential contaminants such as hormones, allergens, and pesticides.

In addition to these industrial-scale applications, there are hints that handheld devices might become available that can detect food contaminants in real time. Many of these work by using various biosensors. For example, the Chinese edition of the *Wall Street Journal* recently reported the use of chopsticks to check for contaminated cooking oil. One group of researchers, writing in the *Proceedings of the National Academy of Sciences,* discussed a smartphone app that could use sensors to detect gases associated with food spoilage.[19] Other scientists are developing handheld sensors that people can use at home to identify bacteria or other food contaminates.[20] This is important because, despite the publicity that often surrounds food recalls by large food companies, the reality is that almost 20 percent of outbreaks occur because of improper food handling at home, and almost 60 percent of illnesses are traced to restaurant meals.[21] Being able to easily detect, for ourselves, the safety of food before we eat it would represent a remarkable breakthrough.

Bailey is a bit skeptical of some of these newer technologies. He thinks they are not likely to be commercially applicable, at least before 2020. He said, "If you don't want absolute precision, then you can get closer to real time. If the standard is absolute zero, then real-time detection is problematic." Still, he thinks such devices might, in the short run, be useful in identifying whether a product is grossly contaminated.

If we can't yet easily test food safety ourselves, we will have to trust those who sell us our food. To take a closer look at food sellers, I got in touch with the biggest around: Frank Yiannas, Walmart's vice president for food safety.[22] Yiannas is a rock star in the world of food safety. He's received numerous awards, written a couple books on the subject, was the president of the International Association for Food Protection, and was the director of safety and health for the Walt Disney World Company before he joined Walmart.

I started by asking about the size of Walmart. More than 120 million Americans (more than a third of the U.S. population) shop at Walmart every week. Does the sheer scale of the operation make the U.S. food system riskier? If Walmart has an outbreak, multitudes would be sickened. Yiannas replied: "One out of every four dollars spent on food are spent at a Walmart. We can make a big difference. Large organizations like Walmart result in a safer food system." He points out that when Walmart makes a change, it affects the whole system. Sure, smaller companies might have outbreaks that affect fewer people, but when lots of small companies are having lots of small outbreaks, the problem is more widespread. A downside to small companies, said Yiannas, is that they can't easily invest in improving the system as a whole. While Walmart often attracts negative attention because of its size and scale (e.g., Do they pay workers fairly? Do they hurt local mom-and-pop businesses?), at least in the world of food safety, their size has significant benefits for its customers, and, as I'll soon discuss, even for non-customers.

The discussion reminded me of some recent research by the economist Marc Bellemare at the University of Minnesota.[23] Since 2005 the number of small food companies has exploded because of the expansion of farmers' markets. Farmers' markets are fun and they provide an opportunity to meet your farmer neighbors and buy tasty products. But is the food sold there safe?

Bellemare investigated the change in the number of farmers' markets in each state during a seven-year period and correlated it with data from the CDC on food-borne illness outbreaks. He found a surprising result that persisted even after controlling for a host of confounding factors: more famers' markets led to more food poisoning. Bellemare's estimates suggest that a 1 percent increase in the number of farmers' markets in given location led to a 0.7 percent increase in the number of food-borne illness outbreaks in that location. He projected that a doubling of the number of farmers' markets in a state would result in an annual cost of $900,000—the combined cost of doctor's visits, medicine, and lost work days—in additional cases of food-borne illness. If the findings seem a bit farfetched, note that other researchers who have tested directly for bacterial contamination in vegetables and poultry have found greater contamination in farmers' markets than in supermarkets.[24] None of this is meant to besmirch farmers' markets, but there is likely a tradeoff between having smaller food sellers with more food safety outbreaks and having large food sellers with fewer food safety outbreaks. In aggregate, it is quite possible that the larger-safer system results in fewer total foodborne illnesses than the smaller-riskier system.

I first met Yiannas a couple years ago at the Global Food Safety Conference, an international meeting of food manufacturers and retailers who created the Global Food Safety Initiative (GFSI). Yiannas was vice chair of the organization, and his presence there illustrates the kind of impact a company like Walmart can have on the safety of the food system. In 2008 Walmart became the first U.S. retailer to require its suppliers to be certified in at least one of the GFSI benchmark standards. The GFSI project aims to improve the safety of food by harmonizing a host of existing international food safety standards so that food companies can adopt science-based standards without having to worry about conflicting company and governmental regulations that result in higher food costs, barriers to entry, and trade prohibitions. Two years later Walmart began a beef safety initiative that requires its suppliers to adopt interventions and prove that they have seen a one-hundred thousandfold reduction in certain pathogens.[25] In 2014, working with the U.S. Department of Agriculture and the CDC, Walmart similarly required poultry suppliers to reduce salmonella by 99.99 percent.[26]

Yiannas said after the beef safety initiative, the number of recalls issued by Walmart's suppliers fell by half. That's not just recalls of product sent to Walmart but also to other retailers by those suppliers, which means Walmart's rules affected the safety of eating hamburger even for customers who bought their beef elsewhere. When Walmart required its suppliers to adopt GFSI standards, many made significant investments and adopted more systematic and transparent food safety practices.[27] Again, these are food companies that supply not only Walmart but other retailers as well.

Yiannas calls it "retail regulation" and credits it for much of the improvement in food safety of recent decades. While he sees a role for the public sector in the provision of safe food, Yiannas said that government "regulations are the lowest common denominator." The standards of private companies with a reputation on the line are often much higher than what the federal government requires. To make his point Yiannas noted that HACCP was developed by a private company, Pillsbury, which created the process for some work it was doing for the National Aeronautics and Space Administration in the 1960s, and the company soon adopted the program for its own product lines. Many food companies voluntarily adopted HACCP well before it became mandatory for U.S. meat companies in the 1990s. GFSI (which incorporates HACCP) isn't a government regulation but a private-sector initiative that has helped improve global food safety by incorporating it in international business standards like those promulgated by the International Organization for Standardization.

Efforts like HACCP, GFSI, and Walmart's beef safety initiative show that food safety advances are created through improvements in processes, methods, and monitoring, as well as gee-whiz technological innovations. Sometimes, breakthroughs are made when the two combine, and new technologies facilitate the development of new monitoring processes. Yiannas described exactly how the revolution in information technology has enabled Walmart to help make safer dinners.

Most Walmart shoppers have probably seen hot, ready-to-eat rotisserie chickens for sale near the deli counter or checkout line. I've occasionally picked up a rotisserie chicken on the way home from work when I didn't have time to cook from scratch. The rotisserie chickens are surprisingly good and remarkably inexpensive. I'm often surprised to find that the already seasoned and cooked rotisserie chicken is no more expensive than buying the whole bird raw. As a consumer I've come to expect Walmart's rotisserie chicken to be a bit like the ideal porridge in "Goldilocks and the Three Bears": not to hot (or overcooked), not too cold (or undercooked), but just right. The challenge for Walmart is that it has to have just enough chickens ready at just the right time, and while overcooked chicken might cause a complaint, an undercooked one could be fatal. And Walmart has to do it right in not just one location but in the more than forty-five hundred locations across the United States, each with different managers and employees.

How can Walmart ensure all those chickens in all those locations are safe to eat? Well, employees have rules to follow on cooking times and temperatures. But how do we know they did their job? Government health inspectors stop by from time to time to check. According to Yiannas, in one month health inspectors across the country checked the temperature of Walmart chickens ten times. Independent, third-party auditors hired by Walmart measured the cooking temperatures one hundred times that month. A tenfold increase in safety checks is good, but hardly assures me that the chicken I put in my basket was cooked properly.

To address this problem Walmart turned to the power of information technology and Big Data.[28] Now all stores are equipped with new handheld sensors that are used to check cooking temperatures of every single batch. The sensors automatically record and send the information to the web in real time. During the month that health inspectors checked Walmart chickens ten times, the company recorded 1.4 million temperature checks. Whereas earlier inspection methods relied on taking a small sample of readings to check for compliance, Yiannas said the new approach is "N = all." In other words, Walmart employees check every single chicken. Moreover, Walmart no longer has to wait on a report from an inspector or third-party auditor to learn the outcomes. Yiannas can check at any time during the day to see which stores are doing what they should to meet food safety standards. The troves of data can be exploited to find out which stores, which equipment, and which employees are doing better. Perhaps most important, it might just stop you and me from walking out the door with an undercooked chicken.

This chapter only mentioned a few of the hundreds of the technological innovations in the food safety arena that may ultimately improve public health. One challenge mentioned by everyone I talked to in researching this chapter is that it's hard for food retailers to make more money by making safer food. It's just something we consumers have come to expect from our food system. One reason we find deaths from a food-borne illness outbreak so shocking is that they are so rare. We have high expectations, which means the risks for a food company that makes a mistake are potentially huge. So, while food safety might not earn companies a premium, it is economically important. Improving food safety acts as an insurance policy for the reputation of the company's brand.

As a result scientists, entrepreneurs, and food companies are working on new ways to make our food even safer. Nanotechnology is being used to create better food packaging, which can even signal when the contents are spoiled. Nanotechnology preservatives, because they're so small, might even be able to compete with the bugs at their own level, becoming activated when pathogens are detected. Farmers and ranchers are also taking up the challenge of food safety. Vaccines that prevent *E. coli* in beef cattle are now available, and probiotic animal feeds reduce the presence of pathogens by enabling healthy, helpful bacteria to outcompete the harmful ones. Livestock breeders are working to identify strains of cattle, hogs, and poultry that are less susceptible to hosting harmful human pathogens. If all that fails, new DNA testing technology might one day be able to identify the grocery store, meatpacker, feedlot, and rancher that sold the bite of hamburger that made you sick.

11

The Case for Food and Agricultural Innovation

n 1970 Jules Billard penned a feature article for *National Geographic* entitled "The Revolution in American Agriculture" with the subheading, "More Food for Our Multiplying Millions."[1] Forty-four years later, that same publication put out a special issue on the future of food; the lead article, by Jonathan Foley, asked, "Where will we find enough food for 9 billion?"[2] The times have changed, but our biggest concern about food has not.

An illustration by James Blair captures Billard's forty-five-year-old imagining of our food future with a decidedly space-age farmhouse on a hilltop watched over by a businessman-farmer from a "bubble-topped control tower . . . with a computer, weather reports, and a farm-price ticker tape." His fields extend to the distant horizon, interrupted only by the skyscrapers of a nearby city. The fields are bordered by "remote-controlled tiller-combine glides" that allow the farmer to plant, till, and harvest the fields without tractors disturbing and compacting the soil. Automated helicopters apply fertilizers and pesticides. The grains are automatically transported to feed mills and eventually to cattle housed in high-rise enclosures reminiscent of parking garages and designed to "conserve ground space." Tubes alongside the cattle enclosures flush and process manure for use as fertilizer in distant greenhouses that "provide controlled environments for growing high-value crops such as strawberries, tomatoes, and celery."

Yes, some futurists teeter on the edge of technological utopianism (where is that flying car we were promised in the 1950s?), and today's farms may not have the modern architectural flare depicted by the artist. But the reality is not that far off. Soil sensors, drones, satellite images, soy burgers, contour plowing, efficient irrigation, chicken cages, and mechanical harvesters all were discussed as the future of food nearly five decades ago, and they are now a regular part of farm and food practices on what are larger, more specialized, but still family-owned farms.[3] GPS signals drive today's tractors, and fertilizer applicators and planters distribute their payloads based on digital input from soil sensors and crop consultants. Farmers watch the evolution of crop prices and thunderstorms on their smartphones. Farmers apply livestock waste as fertilizer or use it in anaerobic digesters to create energy for the farm. Drones track crop yields, cattle location, and animal health. Farming innovators are moving high-value crops indoors under blue and red light-emitting diodes (LEDs) that give off precisely the wavelengths the plants need in environments that use recycled water, reduce water losses from evaporation, and prevent pests and thus the need for pesticides.[4]

These sorts of scientific breakthroughs and technological developments have produced incredible efficiency gains that have literally helped provide "more food for our multiplying millions." Data from the U.S. Department of Agriculture show that U.S. crop production now is twice what it was in 1970. That would not necessarily be a good thing if we were using more land, water, pesticides, and labor to achieve that higher level of production. But that's not what happened. Herbicide use has remained steady since 1980, and use of insecticides has fallen 77 percent since 1970, all while average pesticide toxicity has fallen dramatically.[5] Agriculture is using half as much labor and 16 percent less land than in 1970 even though farmers are producing much more food.[6] Agriculture has one of the highest rates of productivity growth of any sector of the U.S. economy. Farmers are not only getting more output but are taking better care of the land. Soil erosion has declined substantially since the 1980s—down more than 40 percent. Farms today are increasingly using cover crops and practicing more no-till farming, thanks in part to biotechnology.[7]

If the statistics are starting to make your eyes glaze over, perhaps it's because I'm deliberately belaboring the point. In many objective ways agriculture has gotten much better, a point so often overlooked in food writing that it is worth belaboring some more.

Animal agricultural production also has increased. The amount of pork produced per sow in the United States has increased a remarkable 2,780 pounds, or 240 percent, since 1970. Because we now get so much more meat from each sow, we need fewer sows. In 1970 there were 5.18 sows for every hundred people in the United States, but today there are fewer than two sows per hundred people.[8]

Each cow and heifer now produces 50 percent more beef than in 1970. Remarkably, about 4 billion more pounds of beef are produced today than in 1970 even though farmers are raising almost 40 million fewer cows and heifers. Higher yields imply more affordable, abundant food than we would have had otherwise, but they also convey important environmental benefits. Compared to the late 1970s, farmers now use 19 percent less feed, 33 percent less land, and 12 percent less water to produce a given quantity of beef while generating 19 percent less manure and creating a carbon footprint that is 16 percent smaller.[9]

U.S. agriculture largely delivered on the hopes of the 1970s to satisfy the growling stomachs of a growing world, primarily through innovation and technological development. Yet, it seems Americans are hardly content. While an abundant food supply sufficient for an expanding population remains a top concern, the 1970 and 2014 stories in *National Geographic* also reveal shifts in the food problems that occupy our attention as well as changes in how we envisions addressing them. The 2014 special edition of *National Geographic* argued that "agriculture is among the greatest contributors to global warming" and the "environmental challenges posed by agriculture are huge, and they'll only become more pressing as we try to meet the growing need for food worldwide."[10] Other articles in that issue worried about corporate control, hunger, deforestation, nutrition, food deserts, waste, and more.

Yet it's not clear whether our cultural food pessimism is warranted. I recently found reason for hope in the most unlikely place. I grabbed a burrito from Chipotle, and written on the back of the bag was a short essay by the Harvard psychologist Steven Pinker entitled "A Two-Minute Case for Optimism."[11] He writes that when you look at the numbers, the world is definitively getting better: "Worldwide, fewer babies die, more children go to school, more people live in democracies, more can afford simple luxuries, fewer get sick, and more live to old age." Given that Chipotle has removed from its menu most ingredients from genetically engineered plants, the reason Pinker gives for this optimism is a bit ironic: "Problems that look hopeless may not be; human ingenuity can chip away at them. We will never have a perfect world, but it's not romantic or naïve to work toward a better one."[12] Normally, it would be foolish to look for philosophical wisdom from fast-food purveyors, but sometimes we stumble upon deep insights. A better world doesn't happen by accident but rather because we work at it. Human ingenuity, innovation, and entrepreneurship are the catalysts that bring about positive change in every area of our lives, including food and agriculture.

It might be hard to imagine that many people would be overtly opposed to scientific developments in food and agriculture. Yet that's often the driving philosophy that permeates our food culture, from commentators on both the political Left and the Right. The perspectives on the Right sometimes stem from religious motives that elevate purity and the notion that we shouldn't try to play God—the idea that we should eat food the way "God made it." The Left takes cues from the likes of Jean-Jacques Rousseau and elevates nature as a pure state unadulterated by bigotry or profiteering. The result is a philosophy of romantic traditionalism that is implicitly opposed to technological progress in food and agriculture.

Take, for example, the famed food activist Alice Waters, whose Berkeley, California, restaurant, Chez Panisse, is often credited with sparking the local foods and farm-to-table movements. They are idealistic movements, popularized by the writings of people like the journalist Michael Pollan, who celebrates the natural. In one interview Waters expressed criticism of the molecular gastronomy movement in which chefs apply principles of food science to create new delicacies for fine dining. "I am so hungry for the taste of the real that I'm just not able to get into that which doesn't feel real to me," she said. "It's a kind of scientific experiment. . . . It's not a kind of way of eating that we need to really live on this planet together." Elsewhere she has argued that we should farm and cook to "let things taste of what they are."[13] It's hard to pick up a cookbook or food magazine without seeing homage paid to "real," "honest," or "authentic" food, labels that typically refer to some vision of natural food unadulterated by big business and modern technologies.

The trouble with this view is that nature is in a constant state of evolution. Nature is not immutable; rather, different natures have existed at different times. Plants have evolved (sometimes with our ancestors' help) into the foods we now eat, and even some of our own human genes were co-opted from other species. If broccoli is real food, when did it stop being novel and become real? As the Princeton ethicist and moral philosopher Peter Singer put it when discussing the use of biotechnology in food, "It is a mistake to place any moral value on what is natural. I mean many things are natural, including racism, sexism, war, and all sorts of diseases that we try to fight all the time. So the argument about [genetically engineered] food being unnatural and therefore wrong oversimplifies this debate."[14]

In a masterful essay in which she makes a plea for what she calls culinary modernism, the historian Rachel Laudan shows that in the past, "real" and "natural" food was often dangerous and inedible. It took cooking, fermenting, and selection to make nature edible. She writes:

> As an historian I cannot accept the account of the past implied by Culinary Luddism, a past sharply divided between good and bad, between the sunny rural days of yore and the gray industrial present. . . . The Luddites' fable of disaster, of a fall from grace, smacks more of wishful thinking than of digging through archives. It gains credence not from scholarship but from evocative dichotomies: fresh and natural versus processed and preserved; local versus global; slow versus fast; artisanal and traditional versus urban and industrial; healthful versus contaminated and fatty. History shows, I believe, that the Luddites have things back to front.[15]

Despite bumper-sticker platitudes urging us to do so, few of us would want to eat like our grandmothers actually did. Yes, today low-wage, low-skilled workers are doing menial labor on farms, people in the developing world are hungry, and the planet has suffered environmental degradation. But that has been true for centuries and millennia. Technological advancement and industrialization have been great food equalizers—freeing peasants and serfs from the demands of the land and letting them eat like the royalty they once served. Laudan correctly observes that "were we able to turn back the clock, as they urge, most of us would be toiling all day in the fields or the kitchen; many of us would be starving. Nostalgia is not what we need."[16] What we need is a recognition of the ability of technology to help solve our food problems along with wisdom about how to ensure against the risks that technology can create.

What are our challenges for the future, and how should we face them? I posed those questions to a representative sample of more than a thousand Americans in a recent survey. I asked what the respondents thought were the food and agriculture issues that concerned them for the future. Of the nine issues I asked people to rank, one clearly rose to the top: having affordable food. Other issues of top concern were the healthiness of food, producing enough food to meet the demands of a growing world population, and finding ways to prevent adverse environmental impacts of food production.[17] We can all probably agree that these are important issues. That we face serious challenges is incontrovertible. The more difficult question is, What do we do about them?

I queried my survey respondents. I asked them to express which of two options they thought were most *effective* in addressing the pressing challenges they had previously identified: adopting a more technological agricultural production system—more innovation, science, and research in crops and food—or adopting a more natural agricultural production system—more local, organic, unprocessed crops and food.

Which do you think most people chose? The "natural" solution won in a landslide, picked by more than three-quarters of respondents. I don't doubt the sincerity of the respondents, but I do question the efficacy of their chosen solution. To be sure, all options ought to be on the table, and we don't have to adopt an either/or mind-set as I forced the survey respondents to do. But how will we know if so-called natural solutions are even working if we don't scrutinize the outcomes with experiments and the scientific method, which are arguably the most reliable processes we have for generating reliable knowledge about our world? Even if we choose to eschew technological progress in some areas of food and agriculture, we ought to at least leave the door open for innovation to address future food problems, even if technology isn't the complete solution.

A more optimistic and hopeful food future is one in which people are empowered to use creativity, intellect, and determined experimentation to solve today's problems and fashion the type of future they desire. It is a future in which scientists and farmers are free to innovate and consumers are free to adopt the innovations (or not). Technological progress is a practical way to meaningfully tackle our food problems, and there are strong ethical reasons to support and perhaps even fund technological development in food.

A key benefit of modern food and agricultural technologies has been the increase in abundance, making food more affordable for the most impoverished in society. Yes, it is important to think about how to make healthy fruits and vegetables more accessible for the inner-city poor and ensuring nutritional diversity for rural people in Africa and India. But limiting access to food technologies often has precisely the opposite effect intended by groups that otherwise work to help the poor. For example, one study by researchers from Wageningen University and the University of California at Berkeley calculated that the now decades-long delay in approving golden rice has resulted in 600,000 to 1.2 million additional cases of blindness.[18]

It probably isn't helpful or accurate to say that certain technologies are *necessary* to feed a growing population (no one likes being told they must do something), but at the same time we need to be honest about the costs and trade-offs of restricting food and agricultural technologies. If we want less productivity-enhancing technology, are we willing to accept more deforestation? More expensive food? Population control? We may not like these trade-offs, but that doesn't make them disappear. You can draw a map without the state of Texas, but that doesn't make it go away.

Peter Singer has said, in relation to food science and technology, that "the general attitude that technology is a bad thing is a mistake." Failing to apply science and technology to our food problems can cost lives. Therefore, Singer says, "We need to be more cautious with our caution." He's argued that "the admittedly very real risks of synthetic biology seem decisively outweighed by the hope that it may enable us to avert a looming environmental catastrophe." Remarkably, for a man who has perhaps done more than any to provide the intellectual underpinnings for the modern animal welfare movement by blurring the lines between animals and humans by elevating the status of animal suffering, Singer argues that "we are solving our problems by using our intelligence and reason—that is very much a human trait. People always talk about what distinguishes us from animals and that will be as good a candidate as any." He sees technology "as an extension of our humanity rather than a loss of it."[19]

Of course, Singer is not the ultimate moral authority, and indeed he is a controversial figure. Rather, the point is that technological development in food and agriculture can help us meet moral imperatives, such as helping the least fortunate among us. Innovation is only as good as the end at which it's aimed and the outcomes it ultimately creates. As history has shown, food and agricultural innovation are an effective means to improve our lot in life. But none of that is to say that technological development is perfect or without adverse consequences.

Technological development is disruptive. New technologies bankrupt existing firms. Mechanization puts people out of work. That's true today, and it was true for the technologies we now take for granted. For example, the primary source of power on the farm was once provided by humans and then their animals. But as the landowners in the 1939 novel *The Grapes of Wrath* warned their tenants, "One man on a tractor can take the place of twelve or fourteen families."

In the early part of the twentieth century, machine power began substituting for animal power and human labor in a big way. In the early 1900s U.S. farms had more than 25 million horses and mules. By 1960 there were only about 3 million. That's 22 million fewer horses and mules now in the United States. To put that number of animals into perspective, it would be akin to removing *all* the present-day beef and dairy cattle from the major livestock states of Texas, California, and Oklahoma.[20] In contrast, there were virtually no tractors in 1910, but fifty years later there were almost 4.5 million.[21]

Changes in farm technology have had a profound effect on employment. Back in 1900 more than a third of jobs were on the farm. Today, only about 1.5 percent of workers are on the farm. The potential for unemployment was immense, but in 1900 the economy was vibrant and robust enough to create new jobs for all those displaced workers. Only 18 percent of workers had a white-collar job, whereas 61 percent of us now have such work.[22] Opportunities—often higher-paying, less risky, less strenuous opportunities—off the farm have more than compensated for the loss of jobs on the farm. I've seen it in my own family. All four of my grandparents began life in families that depended on agriculture for a living. Yet none of my aunts or uncles, siblings, or cousins makes a living by farming. We are teachers, engineers, administrators, economists, musicians, actuaries, salesmen, mathematicians, and more.

Creative destruction is the term the economist Joseph Schumpeter used to describe the healthy dynamism of an entrepreneurial marketplace in which new products and processes continually outcompete the status quo. It's probably a term only an economist could love. Yet history shows that the detrimental effects of destruction have been more than offset by the positive effects of creativity.

That's small condolence if you lost your job as a result of creative destruction. One justification for social insurance programs, such as unemployment benefits and food stamps, and for public funding for education and job training, is that they are intended to help compensate the displaced few so that the many might benefit, and to help people develop job skills more consonant with the emerging technology. Maybe we'll need more such programs if the public is to support widespread technological development. But the hard truth is that new technologies will make some jobs obsolete.

But disruption and job loss are not unique to the development of technology. Few commentators argue that we should do nothing about our food problems. In fact, advocates of the more natural agricultural production systems often call for radical change. For example, the celebrity chef Jamie Oliver had a television show called *Food Revolution* and even created an annual Food Revolution Day. Remarking in an interview about the change he'd like to see in the food system, the best-selling author Michael Pollan said, "We'd need tens of millions more people working on farms to grow food the way I think most people would like to see it grown. We'd also see ourselves spending more money on food, and that's very challenging for a lot of people. So it will take a revolution—not just in how we eat but how we live."[23] Presumably, this revolution would involve dramatic shifts in patterns of employment and lost jobs for the hundreds of thousands of people employed at places like McDonald's, Monsanto, and Tyson. Do these people want to work in farms? Do they have the skills to do so?

Most of the so-called more natural production systems are much more labor intensive and would require people to move from service-sector jobs and training them to milk cows, plant cover crops, and bale hay. Ironically, it's often the well-heeled urbanites who voluntarily seek this sort of job change. Writing about the folks selling their wares at the huge annual trade fair of the Specialty Food Association, one food writer remarked, "Based on the backgrounds of many new companies at the show, selling everything from chia seed cookies to fine cheeses, a prior career in corporate law or investment banking seemed to be the main prerequisite for starting a specialty food company."[24] There's absolutely nothing wrong with these career choices, but they illustrate that disruption and change are a regular part of the economic churn and employment change is a consequence of the futures preferred both by technology advocates and their detractors.

Few people are arguing for the status quo. All plans for the future will involve disruption. The question is, what kind of disruption we going to have? Disruption that leads to more or less affordable food? Disruption that leads to more or less land in agriculture? Disruption that leads to more or less convenience, health, and environmental degradation?

The unemployment rate has changed cyclically during the past century, and while people have gained dramatic increases in leisure time, there doesn't appear to be a strong trend toward greater unemployment.[25] That hasn't stopped social commentators from speculating about a world in which machines make everything and leave little for we humans to do but consume the robot-made wares. It is the anti-Malthusian future: we'll produce so much stuff for so little work that scarcity will be a thing of the past.

An optimistic take on this future world might ask why we need jobs at all. We don't live to work but work to live. If all the things we want become incredibly inexpensive, we'd be freed to do other things we might enjoy. Interestingly, many of the socialist manifestos of yesteryear were premised on the idea that the problems of production had been solved and that state ownership and direction would allow us all to relax a bit and enjoy the fruits of mechanization. Edward Bellamy's 1888 book, *Looking Backward,* envisioned a socialist utopia in which people didn't have to engage in much menial labor but could instead dedicate themselves to philosophy, the arts, and other such pursuits. Ironically, it is capitalism that may be leading to the labor-displacing robots that will give us the time to achieve self-actualization.

The more pessimistic view of this future envisions a sort of *Brave New World* dystopia, where safety and security are prized to the point that people spend their days engaged in trifles. Others worry that the benefits of technological advancement will accrue to only a few owners of capital, leading to high levels of inequality. These fears have some merit, but they aren't unique to agriculture. In fact, agriculture now employs only about 1.5 percent of workers in the United States.[26] To the extent that job loss is a concern in agriculture *it has already happened* and without undue effects on the overall economy. Thus the current concerns about robots' replacing humans are primarily related to the nonagricultural economy.

But many people are worried about a sort of inequality that could arise through corporate involvement in agriculture, and they distrust Big Ag and Big Food, which sometimes bring about new technologies and reap the benefits of those developments. It is, of course, true that many large agribusinesses are involved in food production and distribution (though it is also true that 97 percent of farms and 90 percent of the farm acres in the United States are owned by families, not corporations).[27] If you believe the large agribusinesses are capturing the benefits of technological development, why not buy stock in the companies and get in on the action? However, the research shows that the stocks of agribusinesses have generally underperformed in relation to nonagricultural companies.[28]

While it is no doubt true that many food and agribusiness companies have invested in research that gave rise to patented seeds and technologies that later proved profitable, consumers have also benefited through lower food prices. In the United States we spend a smaller share of our income on food than people do in any other country of the world and spend less of our income on food than at any time in history.[29] One challenge of increasing the regulatory cost and burden of technological development is that it favors the major agribusinesses that have deep pockets and teams of lawyers over small start-ups and university scientists. As I have illustrated with examples in this book, many of the more interesting developments in food and agriculture are not coming from the behemoth corporations. The key isn't to try to keep large agribusinesses out of food and agriculture but rather to make sure that the barriers to entry are low enough that anybody can compete with them.

*   *   *

If technological innovation is an effective and desirable way to address our most pressing food problems, how do we get more of it? How do we create more innovative technologies? One of the surest ways to get more of an output is to plug in more inputs. In this case, that means investing more in research and development. The other approach is to create innovative ways to fund research and development and to make the money work more wisely.

A key contributor to the productivity growth that we've already enjoyed in food and agriculture came about from public investments in research and development. Some of this research is conducted directly by federal agencies. For example, in FY 2016 the U.S. Department of Agriculture has budgeted more than $1.4 billion to spend on its own research through the Agricultural Research Service.[30] The Agricultural Research Service conducts basic and applied research on a host of issues related to crop and animal production, environmental impacts, food safety, and nutrition. The agency says its "job is finding solutions to agricultural problems that affect Americans every day from field to table."[31]

Other public research is done on college campuses all around the country. The historical precedent for current agricultural research funding was set by U.S. Representative Justin Morrill, a Republican from Vermont whose bill created land grant universities. The Morrill Act was signed into law by Abraham Lincoln in 1862. It gave land (primarily because the federal government was short on cash) to states to create colleges and universities that taught "agriculture and the mechanic arts." This and subsequent legislation ultimately led to the creation (or further development) of universities in all corners of the United States, from Cornell University and the University of Florida to UC Berkeley and Washington State University and everywhere in between. Every state and territory of the United States has at least one land grant institution. The Hatch Act of 1877 created agricultural experiment stations (most of which are today administered by the land grant universities) and funded agricultural research in partnership with state governments. Later funding in 1914 created the Cooperative Extension program to bring the new knowledge being created in the agricultural colleges directly to farmers and consumers.

This system of funding for food and agricultural research still exists today. The Department of Agriculture's National Institute for Food and Agriculture (NIFA) allocates federal monies to the land grant institutions for agricultural research and extension programs, based partly on a formula for each state and partly on competitive grants open to any applicant. For FY 2016, the agency had budgeted $1.6 billion.[32] A portion of my salary and research funding from Oklahoma State University comes from the U.S. Department of Agriculture through NIFA and the "Hatch funds" formula, and another portion comes from the State of Oklahoma to the Oklahoma Agricultural Experiment Station. In

2014 the State of Oklahoma appropriated more than $26 million to spend on food and agricultural research by Oklahoma State University and another $29 million to spend on agricultural education and outreach for farmers and communities.[33] Across the country thousands of scientists—agronomists, animal scientists, ecologists, economists, entomologists, geneticists, foresters, hydrologists, range and wildlife specialists, plant pathologists, and veterinarians—are working on innovative scientific solutions to our food and agricultural problems because of the system of public funding for the land grant colleges and universities.

All this begs the question: Is it worth it? Have these investments in food and agricultural research paid off? As it turns out, a large body of research asks just that question, and the answer is a definitive yes. One review of thirty-five studies found that the average estimated rate of return on U.S. public agricultural research was 53 percent. This rate is quite high compared to other investment alternatives. For example, the real rate of return on U.S. Treasuries has historically been about 3 percent. Some researchers have criticized the methods used to calculate the high rates of return on agricultural research, but even after correcting for problems they show a cost-benefit ratio of 32:1.[34] That is, every dollar of public spending on agricultural research yields $32 in benefits for consumers, retailers, processors, farmers, and agribusinesses. These studies only calculate the benefits of productivity gains and do not include whatever environmental or food quality improvements agricultural research has brought about.

Despite all this, the rate of growth in public spending on agricultural research has slowed. Some critics argue that the slowdown in spending has caused a slowdown in agricultural productivity growth. The types of research funding available also have changed. An increasingly larger proportion of federal research dollars has shifted away from productivity-enhancing research toward research on social goals like childhood obesity, climate change, and the economic viability of organic production and small farmers. Moreover, public funding is being redirected toward production and marketing practices that prohibit the use of technologies, such as biotechnology or synthetic fertilizers and pesticides, known to increase productivity. All are important areas for research, but when the overall size of the funding pie hasn't been growing much, adding new goals and mandates effectively reduces the funding available for fundamental research in genetics, crop science, and the like. One potentially positive development (depending on one's perspective) is that private investments in agricultural research appear to be increasing, mainly in areas related to crop seed and biotechnology.[35]

It is tempting, perhaps, to conclude that food prices are low enough and that resources should be directed toward other social goals. Yet productivity-enhancing technological developments are slow in coming, and decisions to cut back on current research and development have long-term consequences that are difficult to reverse. Also adoption of agricultural technology requires investors to risk their capital. When public policies (and public opinion) are generally skeptical of agricultural technologies, and in some cases the technologies are targets of derision and prohibition, the risks increase, reducing the incentive to invest.

Because my employer benefits from public investments in agricultural research, perhaps all this discussion about the returns on agricultural research—while well supported by solid empirical studies—might sound a bit self-serving. The good news is that public funding is not the only way to finance agricultural research. Are there innovative ways to find more resources for agricultural research and apply that research to more focused goals?

One answer may lie in an attempt to solve a pressing food problem faced by Napoleon Bonaparte. European domination required a large army. A hungry army, constantly on the move. Supplying enough food and serving it to the troops before it spoiled was a feat almost as daunting as the Battle of Waterloo. How did Napoleon solve his food conundrum? Did he call in the top French scientific experts or create the French National Academy of Science and award grants to aspiring assistant professors? No. Napoleon solved his problem with a prize. He offered twelve thousand francs to anyone who could improve food preservation methods. It took a little more than a decade, but Nicolas Appert found an answer: boil food and pack it in airtight jars. Canning has been with us ever since.

A modern-day incarnation of Napoleon's idea is the XPrize, which began with $10 million offered to anyone who could privately build a vehicle that would carry at least three people into space twice within two weeks. The winners collected their check in 2004. Today the XPrize Foundation offers numerous prizes to provide technological solutions to diverse challenges related to education in developing countries, health care, landing a robot on the moon, and reducing air pollution. I became aware of the possibilities for food and agriculture when William Masters, a former colleague of mine who is now a professor in the Department of Food and Nutrition Policy at Tufts University, began writing about the concept more than a decade ago as a way to encourage innovation in African agriculture. Set a measurable goal—like improving cassava yields in a Nigerian province by 10 percent within five years—and offer a monetary prize that will be awarded to whoever creates the process or technology to achieve the goal, and you've got yourself a worthwhile contest.

Now there are organizations like AgResults, which has funding from the Bill and Melinda Gates Foundation and the governments of Australia, Canada, the United Kingdom, and the United States and is putting the idea for agricultural prize–funded research into action. AgResults offers prizes to improve farm storage conditions in Kenya, foster adoption of biofortified maize in Zambia, and reduce the prevalence of naturally occurring aflatoxin in maize in Nigeria. In the developed world food companies are also turning to crowdsourcing and prizes for research that helps solve problems. Examples include companies like InnoCentive and NineSigma that, as I wrote this, provide platforms that allow food companies to offer prizes in food processing and packaging, food safety, food taste and texture, and even plant regeneration. Prizes motivate people to direct their own time and resources toward the problem identified, and if numerous people compete for the prize, the amount of intellectual investment in the problem can far exceed the value of the prize.

Many nonprofit organizations also fund or conduct food and agricultural research. One of the most well-known examples is the Bill and Melinda Gates Foundation, which gave $3.6 billion in grants in 2013, about $390 million of which went toward agricultural development.[36] Other nonprofits employ their own scientists and conduct their own research. One example in my area is the Noble Foundation.[37]

Lloyd Noble was born the son of a small hardware store owner in Indian Territory in 1896. After a few years as a schoolteacher and a tour of duty in the navy, he went home to speculate for oil. He struck it rich. Because he had spent his lifetime among farmers in what became Oklahoma, Noble developed the belief that the health of the soil is the key to prosperity and sustainability. When he died in 1945, much of his fortune went to the foundation he had established in his father's name to improve agricultural sustainability. The Samuel Roberts Noble Foundation is today the largest independent private agricultural and plant science research institution in the United States. The foundation assists farmers and ranchers and conducts research to improve agricultural productivity.

Although other agricultural research organizations focus on fundamental research, the foundation's CEO, Bill Buckner, the former CEO of Bayer CropScience, told me that "no other organizations [are] exactly like ours" in that the Noble Foundation focuses its efforts on some core areas impacting farmers, particularly on crops, grasses, and grazing techniques that are germane to the farmers in their area without being spread too thin among many constituencies—as is often the case for public institutions.[38] Yet even though the foundation spends about $60 million and has more than $1 billion in assets, Buckner says it is constrained by the "high cost of biotech research (primarily biosafety and field testing) for low-value plant science research" because of the regulatory requirements for conducting that sort of science.[39]

*   *   *

We face some serious challenges ahead. It will take more than what we did in the past, and that's what this book has been about. I've given just a small glimpse of the exciting developments in food and agriculture. Many more entrepreneurs and scientists are at work on new probiotics and food safety innovations, on new packaging and nanotechnology that will prevent and signal food spoilage, on animal and plant breeding to reduce carbon emissions, on food engineering to increase shelf life and reduce waste, on nutrition and obesity prevention, on farming practices to reduce environmental impacts, and much more. I have no idea whether the particular products and technologies will ever make it to our farms and kitchens. But that's not really the point. The point is the process. Experimentation and innovation are what will ultimately help address our food problems. If we'll let them.

Notes

Chapter 1: Overcoming Nature

1. R. W. Emerson, "Farming," in *The Essential Writings of Ralph Waldo Emerson,* ed. B. Atkinson (New York: Modern Library, 2000). 679-680. The article was first published as a chapter in *Society and Solitude* under the title "The Man with the Hoe" and was delivered at the Middlesex Agricultural Society during the cattle show on September 29, 1858.

2. The speech is printed in "Inaugural Address of the President," *The Chemical News* 78, no. 2024 (September 9, 1898): 125.

3. Based on my calculations using information from Economic Research Service, U.S. Department of Agriculture, table 1, "Price Indices and Implicit Quantities of Farm Output and Inputs for the United States, 1948–2011," September 27, 2013, http://www.ers.usda.gov/data-products/agricultural-productivity-in-the-us.aspx#28247.

4. R. Bailey, "Never Right, But Never in Doubt," Reason.com, May 5, 2009, http://reason.com/archives/2009/05/05/never-right-but-never-in-doubt.

5. P. Gerland et al., "World Population Stabilization Unlikely This Century," *Science,* October 10, 2014, 234–37.

6. A. H. Ehrlich and P. R. Ehrlich, "Collapse: What Is Happening to Our Chances?" blog entry at Millenium Alliance for Humanity and Biosphere, January 16, 2014, http://mahb.stanford.edu/blog/collapse-whats-happening-to-our-chances/.

7. R. Morgan Griffin, "Obesity Epidemic 'Astronomical,'" *WebMD,* n.d., http://www.webmd.com/diet/features/obesity-epidemic-astronomical.

8. M. Pollan, *The Omnivore's Dilemma: A Natural History of Four Meals* (New York: Penguin, 2006), 2.

9.   C. L. Ogden et al., "Prevalence of Childhood and Adult Obesity in the United States, 2011–2012," *JAMA* 311, no. 8 (2014): 806–14, http://jama.jamanetwork.com/article.aspx?articleid=1832542.

10. Even Borlaug himself has said as much. Writing in 2000: "This is not to say that the Green Revolution is over. Increases in crop management productivity can be made all along the line: in tillage, water use, fertilization, weed and pest control, and harvesting. However, for the genetic improvement of food crops to continue at a pace sufficient to meet the needs of the 8.3 billion people projected to be on this planet at the end of the quarter century, both conventional technology and biotechnology are needed." N. E. Borlaug, "Ending world hunger. The promise of biotechnology and the threat of antiscience zealotry." *Plant Physiology,* 124 (2000), 487–490.

11. "McAfee, McArdle, and Ohanian on the Future of Work," *EconTalk,* hosted by Russ Roberts, June 2, 2014, http://www.econtalk.org/archives/2014/06/mcafee_mcardle.html#highlights.

12. See "The Locavore's Dilemma," chap. 9 in my 2013 book, *The Food Police* (New York: Crown), for the substance of this critique. See also P. Desrochers and H. Shimizu. *The Locavore's Dilemma: In Praise of the 10,000-mile Diet.* (NewYork: Public Affairs, 2012).

13. R. Laudan, *Cuisine & Empire: Cooking in World History* (Berkeley: University of California Press, 2013), 9-10.

Chapter 2: The Price of Happy Hens

1.   *Scientific Report of the 2015 Dietary Guidelines Advisory Committee: Advisory Report to the Secretary of Health and Human Services and the Secretary of Agriculture, February 2015,* pt. D, chap. 1, p. 17, http://www.health.gov/dietaryguidelines/2015-scientific-report/PDFs/Scientific-Report-of-the-2015-Dietary-Guidelines-Advisory-Committee.pdf.

2.   Some competitors have challenged Hampton Creek's use of the word "mayo" arguing that the label violates FDA standards of identity rules that define mayonnaise as containing egg. The FDA has sent Hampton Creek a warning letter about the matter, but it remains unclear as to whether the name will ultimately be changed.

3.   National Agricultural Statistics Service, U.S. Department of Agriculture, *Chickens and Eggs: 2014 Summary,* February 2015, http://www.nass.usda.gov/Publications/Todays_Reports/reports/lyegan15.pdf.

4.   Per-capita consumption is from Office of the Chief Economist, U.S. Department of Agriculture, "US Egg Supply and Use," *World Agricultural Supply and Demand Estimates,* March 10, 2015, http://www.usda.gov/oce/commodity/wasde/latest.pdf.

5. F. B. Norwood and J. L. Lusk, *Compassion by the Pound: The Economics of Farm Animal Welfare* (Oxford: Oxford University Press, 2011).

6. See data in J. B. Chang, J. L. Lusk, and F. B. Norwood, "The Price of Happy Hens: A Hedonic Analysis of Retail Egg Prices," *Journal of Agricultural and Resource Economics* 35 (2010): 406–23, and J. L. Lusk, "The Effect of Proposition 2 on the Demand for Eggs in California," *Journal of Agricultural & Food Industrial Organization* 8, no. 1 (2010).

7. Tom Silva, telephone interview by author, April 2, 2015.

8. S. Strom, "They're Going to Wish They All Could Be California Hens," *The New York Times,* March 4, 2014, A1, http://www.nytimes.com/2014/03/04/business/theyre-going-to-wish-they-all-could-be-california-hens.html?_r=0.

9. For discussion and references to other studies that look at cage versus cage-free systems, see Norwood and Lusk, *Compassion by the Pound.* The Coalition for Sustainable Egg Supply recently undertook a high-quality comparison of the different systems, the results of which are available at http://www2.sustainableeggcoalition.org/final-results.

10. The catalog is available online at http://www.williams-sonoma.com/shop/agrarian-garden/agrarian-garden-chicken-coops/.

11. "Idaho Confirms Avian Flu in Wild Ducks, Backyard Chicken," *Meatingplace.com,* January 22, 2015, http://www.meatingplace.com/Industry/News/Details/55832?loginSuccess. See also several posts on backyard chickens at *Meatingplace.com* by Yvonne Vizzier Thaxton, director of the Center for Food Animal Wellbeing at the University of Arkansas, http://www.meatingplace.com/Industry/Blogs/Details/46182.

12. S. Strom, "Brought by Wild Neighbors, a Deadly Flu Attacks Turkey Flock," *The New York Times,* April 10, 2015, B1, http://www.nytimes.com/2015/04/10/business/brought-by-wild-neighbors-a-deadly-flu-attacks-turkey-flocks.html?_r=1.

13. Ian Duncan, who holds the emeritus chair in Animal Welfare at the University of Guelph, graciously provided me with some historical context for the development of the enriched colony cages.

14. J. R. Bareham, "A Comparison of the Behaviour and Production of Laying Hens in Experimental and Conventional Battery Cages," *Applied Animal Ethology* 2 (1976): 291–303.

15. Duncan mentioned researchers working in the late 1970s and early 1980s like Brantas, Wegner, and Wennrich.

16. M. C. Appleby, "The Edinburgh Modified Cage: Effects of Group Size and Space Allowance on Brown Laying Hens," *Journal of Applied Poultry Research* 7 (1998): 152–61; M. C. Appleby and B. O. Hughes, "The Edinburgh Modified Cage for Laying Hens," *British Poultry Science* 36 (1995): 707–18.

17. The enriched colony cage with sixty hens provides a bit more than 48 square feet of total space; a king-size mattress is about 42 square feet.

18. Some images from one of the largest manufacturers of enriched colony cages are posted at http://bigdutchmanusa.com/avech/.

19. See http://www.humanesociety.org/assets/pdfs/farm/battery_cage _agreement_fact.pdf and http://www.foodsafetynews.com/2011/07 /fragile-support-for-federal-guidelines-on-hen-cages/#.VS6Zd_nF 98E. Several other animal advocacy groups supported this change as well; see "The Hens Need Your Help," ca. 2012, http://farmsanc tuary.org/wp-content/uploads/2012/07/FAP_FarmSanctuary_Flier _01_12_HRez.pdf.

20. Watch the JS West and Companies' hen cams at http://www.jswest .com/index.php/component/content/article/66 or http://www.jswest .com/index.php/component/content/article/118.

21. R. W. Prickett, F. B. Norwood, and J. L. Lusk, "Consumer Preferences for Farm Animal Welfare: Results from a Telephone Survey of US Households," *Animal Welfare* 19, no. 3 (2010): 335–47.

22. See table 1 in R. M. De Mol et al., "A Computer Model for Welfare Assessment of Poultry Production Systems for Laying Hens," *NJAS-Wageningen Journal of Life Sciences* 54, no. 2 (2006): 157–68.

23. The Rondeel system is described at http://www.rondeel.org/uk/the -system/ and in an elaborate brochure, "Laying Hen Husbandry," available at http://www.rondeel.org/public/files/Houden%20van% 20Hennen%20-%20EN.pdf.

24. For Dutch egg prices see http://www.ah.nl/zoeken?rq=ei (prices as of April 15, 2015). Prices in Euros were converted to US Dollars given an exchange rate of 1 Euro = 1.12 Dollars.

25. R. Schmalensee et al., "An Interim Evaluation of Sulfur Dioxide Emissions Trading," *Journal of Economic Perspectives* 12 (1998): 66.

26. R. N. Stavins, "What Can We Learn from the Grand Policy Experiment? Lessons from S02 Allowance Trading," *Journal of Economic Perspectives* 12 (1998): 84.

27. World Bank, "State of the Carbon Market 2007," press release, May 2, 2007, http://web.worldbank.org/WBSITE/EXTERNAL/NEWS /0,contentMDK:21319772~menuPK:34463~pagePK:34370~piPK: 34424~theSitePK:4607,00.html.

28. "Reduce Your Individual Carbon Footprint," Carbonfund.org, http://www.carbonfund.org/individuals; L. Gagelman, "The Carbon Footprint of Beef, Grass-Finished Beef, Other Meats, and Carbon Offsets," master's thesis, Department of Agricultural Economics, Oklahoma State University, 2015. The price given for offsetting the flight is as of April 20, 2015.

29. J. L. Lusk, "The Market for Animal Welfare," *Agriculture and Human Values* 28, no. 4 (2011): 561–75.

30. For descriptions of such models see, for example, R. Botreau, et al., "Aggregation of Measures to Produce an Overall Assessment of Animal Welfare, Part 1: A Review of Existing Methods," *Animal* 1 (2007):1179–87; R. Botreau et al., "Aggregation of Measures to Produce an Overall Assessment of Animal Welfare, Part 2: Analysis of Constraints," *Animal* 1 (2007): 1188–97; M.B.M. Bracke et al., "Decision Support System for Overall Welfare Assessment in Pregnant Sows A: Model Structure and Weighting Procedure," *Journal of Animal Science* 80 (2002): 1819–34; M.B.M. Bracke et al., "Decision Support System for Overall Welfare Assessment in Pregnant Sows B: Validation by Expert Opinion," *Journal of Animal Science* 80 (2002): 1835–45; R. M. De Mol et al., "A Computer Model for Welfare Assessment of Poultry Production Systems for Laying Hens," *Netherlands Journal of Agricultural Science* 54 (2006): 157–68.

31. The number of Facebook likes is given as of April 23, 2015 (https://www.facebook.com/humanesociety). For donations and support, see HSUS, "Celebrating Animals, Confronting Cruelty: Annual Report 2014," n.d., http://www.humanesociety.org/assets/pdfs/about/2014-hsus-annual-report.pdf.

32. "Ballot Watch: Proposition 2: Standards for Confining Farm Animals," *The Sacramento Bee*, September 27, 2008.

Chapter 3: Hewlett Packard with a Side of Fries

1. "Marginalized," *Lapham's Quarterly* 5, no. 2 (Spring 2012), http://laphamsquarterly.org/communication/charts-graphs/marginalized.

2. Hod Lipson, telephone interview by author, April 30, 2015. See also H. Lipson and M. Kurman, *Fabricated: The New World of 3D Printing* (Indianapolis: Wiley, 2013); D. Periard et al., "Printing Food," *Proceedings of the 18th Solid Freeform Fabrication Symposium*, August 2007, 564–74; and J. I. Lipton et al., "Multi-Material Food Printing with Complex Internal Structure Suitable for Conventional Post-Processing," 21st Solid Freeform Fabrication Symposium, 2010, Austin, TX.

3. Jeffrey Lipton and Hod Lipson, "Adventures in Printing Food," *IEEE Spectrum*, May 31, 2013, http://spectrum.ieee.org/consumer-electronics/gadgets/adventures-in-printing-food.

4. H. Ledford, "Foodies Embrace 3D-Printed Cuisine," *Nature News*, April 20, 2015, http://www.nature.com/news/foodies-embrace-3d-printed-cuisine-1.17358.

5. Mark Oleynik, telephone interview by author, May 7, 2015. See also "Domestic Automation. Robochef Gets Cooking," *The Economist,* April 18, 2015, http://www.psfk.com/2015/04/robot-chef -moley-robotics-shadow-robot.html, and "Robotic Chef Can Cook Michelin Star Food in Your Kitchen by Mimicking World's Best Cooks," *IBTimes UK,* April 14, 2015, https://www.youtube.com /watch?v=QDprrrEdomM.

6. L. P. Smith, Shu Wen Ng, and B. M. Popkin, "Trends in US Home Food Preparation and Consumption: Analysis of National Nutrition Surveys and Time Use Studies from 1965–1966 to 2007–2008," *Nutrition Journal* 12 (2013): 45.

7. D. Cutler, E. Glaeser, and J. Shapiro, "Why Have Americans Become More Obese?" *Journal of Economic Perspectives* 17 (2003): 93–18; S. M. Bianchi et al., "Is Anyone Doing the Housework? Trends in the Gender Division of Household Labor," *Social Forces* 79, no. 1 (September 2000): 191–228.

8. P. R. Liegey, "Hedonic Quality Adjustment Methods for Microwave Ovens in the U.S. CPI," October 16, 2001, Bureau of Labor Statistics, U.S. Department of Labor, http://www.bls.gov/cpi/cpimwo.htm.

9. In 1979 fewer than 10 percent of U.S. households had a microwave. By the end of the 1980s, more than 80 percent had one. See J. F. Guenthner, Biing-Hwan Lin, and A. E. Levi, "The Influence of Microwave Ovens on the Demand for Fresh and Frozen Potatoes," *Journal of Food Distribution Research,* September 1991, 45–52, http://ageconsearch .umn.edu/bitstream/27501/1/22030045.pdf; D. Schardt, "It Was Forty Years Ago Today," *Nutrition Action Health Letter,* January– February 2011, http://cspinet.org/nah/articles/40yearsago.html.

Chapter 4: Synthetic Biology

1. For a great history of these developments see Maureen Ogle's book *Ambitious Brew: The Story of American Beer* (Orlando, FL: Harcourt, 2006).

2. Jeffrey Kahn, "How Beer Gave Us Civilization," *The New York Times,* March 17, 2013, SR9, http://www.nytimes.com/2013/03/17/opin ion/sunday/how-beer-gave-us-civilization.html?_r=0.

3. Adam Cole, "Shall I Encode Thee in DNA? Sonnets Stored on Double Helix," *NPR.org,* January 24, 2013, http://www.npr.org /2013/01/24/170082404/shall-i-encode-thee-in-dna-sonnets-stored -on-double-helix.

4. Mohammad Katouli, "Population Structure of Gut *Escherichia coli* and Its Role in Development of Extra-intestinal Infections," *Iranian*

Journal of Microbiology 2, no. 2 (2010): 59–72, http://www.ncbi.nlm
.nih.gov/pmc/articles/PMC3279776/.

5. I based this estimate on figures in Centers for Disease Control and
Prevention, "National Diabetes Fact Sheet," 2011, http://www.cdc
.gov/diabetes/pubs/pdf/ndfs_2011.pdf.

6. For a discussion and history see S. B. Leichter, "The Business of Insulin: The Relationship Between Innovation and Economics," *Clinical
Diabetes* 21, no. 1 (January 2003), http://clinical.diabetesjournals.
org/content/21/1/40.full, or A. teuscher, *Insulin: A Voice for Choice*
(Basel: S. Karger, 2007), https://www.karger.com/produktedb/kata
logteile/isbn3_8055/_83/_53/insulin_02.pdf.

7. The 80 percent figure comes from M. E. Johnson and A. E. Lucey,
"Major Technological Advances and Trends in Cheese," *Journal of
Dairy Science* 89, no. 4 (April 2006), http://www.ncbi.nlm.nih.gov
/pubmed/16537950.

8. Evolva, "Evolva to Acquire Allyix," press release, November 18, 2014,
http://www.evolva.com/press-release/evolva-to-acquire-allylix/.

9. Many of these companies have been featured in *The Atlantic,* National Public Radio, *The New York Times, The Washington Post,* and
others. See, for example, J. Barthwaite, "The Rise of GMOs: Beyond
Synthetic Biology," *The Atlantic,* September 25, 2014, http://m.the
atlantic.com/technology/archive/2014/09/beyond-gmos-the-rise
-of-synthetic-biology/380770/; D. Charles, "Who Made That Flavor?
Maybe a Genetically Altered Microbe," *NPR.org,* December 4, 2014,
http://www.npr.org/blogs/thesalt/2014/12/04/368001548/who
-made-that-flavor-maybe-a-genetically-altered-microbe; E. Barclay,
"GMOs Are Old Hat: Synthetically Modified Food Is the New
Frontier," *NPR.org,* October 3, 2014, http://www.npr.org/blogs/the
salt/2014/10/03/353024980/gmos-are-old-hat-synthetically-mod
ified-food-is-the-new-frontier; G. Harman, "Technology Is Ready
for Synthetic Foods. Are You?" *The Guardian,* December 17, 2014,
http://www.theguardian.com/sustainable-business/2014/dec/17
/synthetic-foods-technology-labeling-gmo; T. C. Nguyen, "Animal
Lovers Use Biotech to Develop Milk Made by Man Instead of a
Cow," *The Washington Post,* July 21, 2014, http://www.washington
post.com/national/health-science/animal-lovers-use-biotech-to-de
velp-milk-made-by-man-instead-of-a-cow/2014/07/21/c79e4ea6
-0d07-11e4-b8e5-d0de80767fc2_story.html; W. Herkewitz, "Scientists Create Synthetic Yeast, and Open the Door to the Future
of Beer," *PM,* March 27, 2014, http://www.popularmechanics.com
/science/health/genetics/scientists-create-synthetic-yeasts-and-open
-the-door-to-the-future-of-beer-16637455; A. Pollack, "What's That

Smell? Exotic Scents Made from Re-engineered Yeast," *The New York Times,* October 20, 2013, http://www.nytimes.com/2013/10/21/busi ness/whats-that-smell-exotic-scents-made-from-re-engineered-yeast. html.

10. The address for the competition website is http://igem.org/. The site describes the competition, includes links to team projects over the years, and lists previous competition winners.

11. C. L. Ogden et al., "Prevalence of Childhood and Adult Obesity in the United States, 2011–2012," *JAMA* 311, no. 8 (2014): 806–14, http:// jama.jamanetwork.com/article.aspx?articleid=1832542; Centers for Disease Control and Preventions, "Obesity and Overweight," *Fast-Stats,* 2014, http://www.cdc.gov/nchs/fastats/obesity-overweight.htm.

12. C. Cawley and C. Meyerhoefer, "The Medical Care Costs of Obesity: An Instrumental Variables Approach," *Journal of Health Economics* 31, no. 1 (January 2012): 219–30, http://www.sciencedirect.com/science /article/pii/S0167629611001366.

13. K. Cacazza et al., "Myths, Presumptions, and Facts About Obesity," *New England Journal of Medicine* 368 (January 31, 2013), http:// www.nejm.org/doi/full/10.1056/NEJMsa1208051.

14. See my discussion of this topic in "The Thin Logic of Fat Taxes," chap. 8, in *Food Police* (Orlando, FL: Crown, 2013) and the references cited therein.

15. Marketdata Enterprises, *The U.S. Weight Loss & Diet Control Market,* 12th ed. (Tampa, FL: March 2013).

16. Paul (Yui Shing) Tse and Marco (Lok Man) So, e-mails to author, January 2015.

17. "Health Facts of Hong Kong, 2015 Edition," http://www.dh.gov.hk /english/statistics/statistics_hs/files/Health_Statistics_pamphlet_E .pdf.

18. Tribute Pharmaceuticals Canada, "What Is Mutaflor?" 2015, http:// mutaflor.ca/what-is-mutaflor/.

19. L. Grozdanov et al., "Analysis of the Genome Structure of the Non-pathogenic Probiotic Escherichia coli Strain Nissle 1917," *Journal of Bacteriology* 186, no. 16 (August 2014): 5432–41, doi:10.1128/ JB.186.16; N. Kamada et al., "Nonpathogenic Escherichia coli Strain Nissle1917 Prevents Murine Acute and Chronic Colitis," *Inflammatory Bowel Diseases* 11, no. 5 (May 2005): 455–63, doi:10.1097/01. MIB.0000158158.

20. Sarah Ritz and Aaron Cohen, telephone interviews by author, January 13, 2015

21. California routinely ranks first out of all fifty states in terms of the value of agricultural production, and UC Davis often appears in the upper echelon of rankings of agricultural universities. Economic Research

Service, U.S. Department of Agriculture, "Farm Income and Wealth Statistics," http://www.ers.usda.gov/data-products/farm-income-and -wealth-statistics/annual-cash-receipts-by-commodity.aspx# .VLVJgCvF98E; "Best Global Universities for Agricultural Science," *U.S. News & World Report,* 2015, http://www.usnews.com/educa- tion/best-global-universities/agricultural-sciences; "Top Agriculture Universities by Rank," *Western Farm Press,* May 14, 2013, http://west ernfarmpress.com/management/top-agriculture-universities-rank.

22. There may be trivial amounts of olive production in other states, but the vast majority is in California; California is the only state for which the U.S. Department of Agriculture tracks data on olive production. US Department of Agriculture, *QuickStats,* http://quickstats.nass .usda.gov/results/8CD4F789-BED1-349C-B651-3A7D13C03C83.

23. D. Pierson, "California Adopts New Olive Oil Standards," *Los Angeles Times,* September 18, 2014, http://www.latimes.com/business/la-fi -olive-oil-20140918-story.html.

24. For details of the project see the team's Wiki page at http://2014 .igem.org/Team:UC_Davis.

Chapter 5: Growing Flintstones

1. A refreshing counterpoint is the evidence-based writing by the econo- mist Emily Oster, *Expecting Better: Why the Conventional Pregnancy Wisdom Is Wrong—and What You Really Need to Know* (New York: Penguin, 2013).

2. H. Levenstein, *Fear of Food: A History of Why We Worry About What We Eat* (Chicago: University of Chicago Press, 2012).

3. R. E. Black et al., "Maternal and Child Undernutrition and Over- weight in Low-Income and Middle-Income Countries," *The Lancet* 382, no. 9890 (August 3, 2013): 427–51.

4. B. de Benoist et al., *Worldwide Prevalence of Aenemia 1993–2005: WHO Global Database on Aenemia* (Geneva: World Health Orga- nization, 2008), http://whqlibdoc.who.int/publications/2008/9789 241596657_eng.pdf.

5. Institute for Health Metrics and Evaluation (IHME), Global Burden of Disease Database, University of Washington, Seattle, 2014,

6. Ibid.; B. de Benoist et al., "Iodine Deficiency in 2007: Global Progress Since 2003," *Food and Nutrition Bulletin* 29, no. 3 (2008): 195–202, http://www.who.int/nutrition/publications/micronutrients/FNB vol29N3sep08.pdf.

7. IHME, Global Burden of Disease Database; Black et al., "Maternal and Child Undernutrition and Overweight"; World Health Organi- zation, *Global Prevalence of Vitamin A Deficiency in Populations at Risk*

1995–2005: WHO Global Database on Vitamin A Deficiency (Geneva: World Health Organization, 2009), http://whqlibdoc.who.int/publi cations/2009/9789241598019_eng.pdf; see reference cited in Justus Wesseler and David Zilberman, "The Economic Power of the Golden Rice Opposition," *Environment and Development Economics,* 19, no. 6 (December 2014), doi:10.1017/S1355770X1300065X.

8. A plant breeder employed by HarvestPlus told me that it is difficult, if not impossible, to use conventional breeding techniques to improve vitamin B content, often because of the lack of genetic variation in expression of the nutrient in different varieties. As I will discuss later, genetic engineering approaches can be (and have been) used to increase vitamin B, although not everyone agrees that this approach is feasible.

9. M. F. Holick and T. C. Chen, "Vitamin D Deficiency: A Worldwide Problem with Health Consequences," *American Journal of Clinical Nutrition* 87, no. 4 (April 2008): 10805–65, http://ajcn.nutrition .org/content/87/4/1080S.short.

10. See J. Wesseler and D. Zilberman, "The Economic Power of the Golden Rice Opposition."

11. S. Horton, H. Alderman, and J. A. Rivera, "Hunger and Malnutrition," Copenhagen Consensus 2008 Challenge Paper, Copenhagen Consensus Center, May 11, 2008. See also: http://www.copenhagen consensus.com/sites/default/files/cc08_results_final_0.pdf

12. H. De Steur et al., "Status and Market Potential of Transgenic Biofortified Crops," *Nature Biotechnology* 33, no. 1(2015): 25–29. 26

13. International Fund for Agricultural Development and United Nations Environment Programme, *Smallholders, Food Security, and the Environment,* 2013, http://www.unep.org/pdf/SmallholderReport_WEB .pdf.

14. H. De Steur et al., "Status and Market Potential."

15. Republic of Mozambique, *Multisectoral Plan for Chronic Malnutrition Reduction in Mozambique 2011–2014* (Maputo, Mozambique: July 2010), http://www.who.int/nutrition/landscape_analysis/Mozambi queNationalstrategyreductionstunting.pdf.

16. For life expectancy (and other statistics) in Mozambique, see World Health Organization statistics at http://www.who.int/countries/moz /en/; comparable statistics from WHO for the United States may be found at http://www.who.int/countries/usa/en/.

17. See Abdul T. A. Naico, "Consumer Preferences of Orange and White-Fleshed Sweet Potato: Results from a Choice Experiment Conducted in Maputo and Gaza Provinces, Mozambique," Master's thesis, Oklahoma State University, May 2009.

18. C. Hotz et al., "A Large-Scale Intervention to Introduce Orange Sweet Potato in Rural Mozambique Increases Vitamin A Intakes Among Children and Women," *British Journal of Nutrition* 108 (2012): 163–76; C. Hotz et al., "Introduction of β-Carotene-Rich Orange Sweet Potato in Rural Uganda Resulted in Increased Vitamin A Intakes Among Children and Women and Improved Vitamin A Status Among Children," *Journal of Nutrition* 142, no. 10 (2012): 1871–80; HarvestPlus, *Disseminating Orange-Fleshed Sweet Potato: Findings from a HarvestPlus Project in Mozambique and Uganda* (Washington, DC: HarvestPlus, 2010).

19. HarvestPlus is run by the International Center for Tropical Agriculture and the International Food Policy Research Institute. Full disclosure: I have served as a paid consultant for HarvestPlus on research related to the design and analysis of consumer research related to acceptance of nutrient-rich staple crops.

20. HarvestPlus has released consumer acceptance studies on Ugandan orange sweet potatoes, Nigerian vitamin A cassava, and Zambian vitamin A maize. See J. V. Meenakshi et al., "Using a Discrete Choice Experiment to Elicit the Demand for a Nutritious Food: Willingness-to-Pay for Orange Maize in Rural Zambia," *Journal of Health Economics* 31 (2012): 62–71; S. Chowdhury et al., "Are Consumers in Developing Countries Willing to Pay More for Micronutrient-Dense Biofortified Foods? Evidence from a Field Experiment in Uganda," *American Journal of Agricultural Economics* 93, no. 1 (2011): 83–97; A. Oparinde et al., "Information and Consumer Willingness to Pay for Biofortified Yellow Cassava: Evidence from Experimental Auctions in Nigeria, *Agricultural Economics,* forthcoming.

21. See A. Oparinde et al., "Consumer Acceptance of Biofortified Iron Beans in Rural Rwanda: Experimental Evidence," HarvestPlus Working Paper No. 18, iMarch 2015.

22. Ibid.

23. UN Food and Agriculture Organization, "Staple Foods: What Do People Eat?" FAO Corporate Document Repository, n.d., http://www.fao.org/docrep/u8480e/u8480e07.htm.

24. Ingo Potrykus, email message to author, November 10, 2014.

25. One of the seminal scientific papers demonstrating that rice could produce beta-carotene was Ye X et al., "Engineering the Provitamin A (Beta-carotene) Biosynthetic Pathway into (Carotenoid-Free) Rice Endosperm," *Science* 287 (2000): 303–5.

26. J. A. Paine et al., "Improving the Nutritional Value of Golden Rice Through Increased Pro-Vitamin A Content," *Nature Biotechnology* 23, no. 4 (2005): 482–87.

27. Guangwen Tang et al., "Golden Rice Is an Effective Source of Vitamin A," *American Journal of Clinical Nutrition* 89 (June 2009): 1776–83; Guangwen Tang et al., "β-Carotene in Golden Rice Is as Good as β-carotene in Oil at Providing Vitamin A to Children," *American Journal of Clinical Nutrition* 96 (September 2012): 658–64.

28. De Steur et al., "Status and Market Potential."

29. Ibid.; E. Kikulwe et al., "A Latent Class Approach to Investigating Developing Country Consumers' Demand for Genetically Modified Staple Food Crops: The Case of GM Banana in Uganda," *Agricultural Economics* 42, no. 5 (2011): 547–60.

30. J. L. Lusk, "Effects of Cheap Talk on Consumer Willingness-to-Pay for Golden Rice," *American Journal of Agricultural Economics* 85, no. 4 (2003): 840–56; J. L. Lusk and A. Rozan, "Consumer Acceptance of Biotechnology and the Role of Second Generation Technologies in the USA and Europe," *TRENDS in Biotechnology* 23, no. 8 (2005): 386–87.

31. See table 2 in J. R. Corrigan et al., "Comparing Open-Ended Choice Experiments and Experimental Auctions: An Application to Golden Rice," *American Journal of Agricultural Economics* 91, no. 3 (2009): 837–53.

32. Wesseler and Zilberman, "Economic Power of the Golden Rice Opposition."

Chapter 6: Farming Precisely

1. J. M. MacDonald, P. Korb, and R. A. Hoppe, *Farm Size and the Organization of U.S. Crop Farming*, ERR-152 (Washington, DC: Economic Research Service, U.S. Department of Agriculture, August 2013), http://www.ers.usda.gov/media/1156726/err152.pdf. According to this publication, "Midpoint acreage is revealed to be a more informative measure of cropland consolidation than either a simple median (in which half of all farms are either larger or smaller) or the simple mean (which is average cropland per farm)." 1 acre = 0.75625 American football fields, including end zones, or 0.9075 football fields, excluding end zones.

2. David Waits, interview by author, Stillwater, Oklahoma, November 18, 2014.

3. M. Oliver, T. Bishop, and B. Marchant, eds., *Precision Agriculture for Sustainability and Environmental Protection* (New York: Routledge, 2013). In the first chapter Oliver writes, "Farmers divided their land into smaller areas, the characteristics of which they knew well. . . . In Britain there is evidence of small fields that were relatively uniform, each of which could be managed as a unit, and that have since been

joined to form much larger fields that are consequently more variable" (5).

4. Despite the romantic appeal of running a small farm, many who try it become somewhat disillusioned. See B. Smith, "Don't Let Your Children Grow Up to Be Farmers," *The New York Times,* August 9, 2014, http://www.nytimes.com/2014/08/10/opinion/sunday/dont-let -your-children-grow-up-to-be-farmers.html; C. Passy, "The New Gentleman Farmer," *The Wall Street Journal,* December 12, 2013, http://www.wsj.com/news/articles/SB10001424052702303997 604579242722533288250.

5. M. Selman and S. Greenhalgh, "Eutrophication: Policies, Actions, and Strategies to Address Nutrient Pollution," *WRI Policy Note,* no. 3 (September 2009), http://www.wri.org/sites/default/files/pdf/eutro phication_policies_actions_and_strategies.pdf.

6. G. Schnitkey, "Crop Budgets, Illinois, 2014," *Farm Business Management,* June 2014, table 2, http://www.farmdoc.illinois.edu/manage /2014_crop_budgets.pdf; L. Tourte and R. Smith, "Sample Production Costs for Wrapped Iceberg Lettuce Sprinkler Refrigerated—40-inch Beds, Central Coast 2010," University of California Cooperative Extension, 2010, http://coststudies.ucdavis.edu/files /2010Lettuce_Wrap_CC.pdf; L. Tourte et al., "Sample Costs to Produce Organic Leaf Lettuce," University of California Cooperative Extension, 2009, http://coststudies.ucdavis.edu/files/lettuceleaf organiccc09.pdf.

7. A large body of economic research has addressed this question. See, for example, B. A. Babcock, "The Effects of Uncertainty on Optimal Nitrogen Applications," *Review of Agricultural Economics* 14, no. 2 (1992), 271–80; G. Sheriff,. "Efficient Waste? Why Farmers Overapply Nutrients and the Implications for Policy Design," *Applied Economic Perspectives and Policy* 27, no. 4 (2005): 542–57; M. Isik and M. Khanna, "Stochastic Technology, Risk Preferences, and Adoption of Site-Specific Technologies," *American Journal of Agricultural Economics* 85, no. 2 (2004): 305–17; N. D. Paulsonand B. A. Babcock, "Readdressing the Fertilizer Problem," *Journal of Agricultural and Resource Economics* 35, no. 3 (2010): 368–84.

8. M. Oliver, "An Overview of Precision Agriculture," in Oliver, Bishop, and Marchant, *Precision Agriculture for Sustainability and Environmental Protection,* 7.

9. Ibid. R. Gebbers and V. I. Adamchuk, "Precision Agriculture and Food Security," *Science,* February 12, 2010, 828–31.

10. R. Bongiovanni and J. Lowenberg-DeBoer, "Precision Agriculture and Sustainability," *Precision Agriculture* 5, no. 4 (2004): 359–87; M. Diacono, P. Rubino, and F. Montemurro, "Precision Nitrogen

Management of Wheat: A Review," *Agronomy for Sustainable Development* 33, no. 1 (2004): 219–41; Oliver, "An Overview of Precision Agriculture," 13.

11. Matt Waits explained the different technologies available. Corn farmers sometimes use a "split planter" approach, that is, they plant eight rows of one hybrid and eight rows of another hybrid in a sixteen-row planter, either to compare yields of the two varieties or to reduce risk through diversification. Another common approach is to load four rows of a twelve-row planter with a non-GMO variety (and the other eight with an insect-resistant variety) to mitigate against development of insect resistance. The newest technology is "multihybrid planting": the farmer loads two hybrids into a new planter that can change (based on a recommendation sent to its onboard computer) which of the two hybrids are planted as the tractor moves through the field.

12. A large literature considers the effect of USDA reports on futures market prices. Two recent examples are M. K. Adjemian, "Quantifying the WASDE Announcement Effect," *American Journal of Agricultural Economics* 94, no. 1 (2012): 238–56; O. Isengildina-Massa, B. Karali, and S. H. Irwin, "When Do the USDA Forecasters Make Mistakes? *Applied Economics* 45, no. 36 (2013): 5086–103.

13. D. Charles, "Big Data Companies Agree: Farmers Should Own Their Information," *NPR.org*, November 16, 2014, http://www.npr.org /blogs/thesalt/2014/11/16/364115200/big-data-companies-agree -farmers-should-own-their-information.

14. "Pointers to the Future: Forecasting the Internet's Impact on Business Is Proving Hard," *The Economist*, October 18, 2014, http://www .economist.com/news/business/21625801-forecasting-internets -impact-business-proving-hard-pointers-future.

15. W. Berry, "Farmland Without Farmers," *TheAtlantic.com*, March 19, 2015, http://m.theatlantic.com/national/archive/2015/03/farmland -without-farmers/388282/.

Chapter 7: Bovine in a Beaker

1. My paternal step-grandfather was Jack Jordan. My biological grandfather, J. R. Lusk, died when my father was a child. Jack married my grandmother well before I was born, making him the only paternal grandfather I ever knew.

2. Some of the source material in this chapter comes from Post's "Meet the New Meat" presentation in Nashville at a PIC Symposium on May 13, 2014.

3. The film's director, Kip Andersen, said, "A lot of us are waking up and realizing we can choose to either support all life on this planet or

kill all life on this planet, simply by virtue of what we eat day in and day out. One way to eat takes life, while another spares as many lives (plant, animal and otherwise) as possible." "Director Kip Andersen Talks New Documentary 'Cowspiracy,'" *TheSource.com*, August 9, 2014.

4. B. Maher, "A-hole in One Shouldn't Be Obama's Game," *HuffPost* (blog), *Huffington Post*, May 25, 2011, http://www.huffingtonpost .com/bill-maher/new-rule-a-hole-in-one-sh_b_259281.html; M. Bittman, "What's Wrong with What We Eat," TED Talk, May 2008, http://www.ted.com/talks/mark_bittman_on_what_s_wrong _with_what_we_eat/transcript; J. McWilliams, "The Myth of Sustainable Meat," *The New York Times*, April 12, 2012; J. McWilliams, *The Modern Savage: Our Unthinking Decision to Eat Animals* (New York: St. Martin's, 2015).

5. *Scientific Report of the 2015 Dietary Guidelines Advisory Committee: Advisory Report to the Secretary of Health and Human Services and the Secretary of Agriculture, February 2015*, http://www.health.gov /dietaryguidelines/2015-scientific-report/PDFs/Scientific-Report-of -the-2015-Dietary-Guidelines-Advisory-Committee.pdf.

6. The bag can be viewed at PETA's CafePress store: http://www.cafe press.com/petastore.309786585.

7. For data on the percentage of the population that is vegetarian or vegan, see the monthly *Food Demand Survey* (*FooDS*), a publication that I run at Oklahoma State University. A summary of the second years' worth of results is available in Lusk, J.L. and S. Murray. "Food Demand Survey: Second Year Summary." Department of Agricultural Economics, Oklahoma State University, April 2015. http:// agecon.okstate.edu/faculty/publications/5048.pdf.ss For a summary about vegetarianism, see J. L. Lusk, "Who Are the Vegetarians?" *Jayson Lusk* (blog), September 9, 2014, http://jaysonlusk.com/blog /2014/9/30/who-are-the-vegetarians. For data on backsliding vegetarians see C. Green, "How Many Former Vegetarians and Vegans Are There?" *faunalytics* (blog), Humane Research Council, December 2, 2014, http://spot.humaneresearch.org/content/how-many-for mer-vegetarians-are-there.

8. Jean-Louis Flandrin and Massimo Montanari, eds., *Food: A Culinary History from Antiquity to the Present*, trans. Albert Sonnenfeld (New York: Columbia University Press, 1999); C. B. Stanford, *The Hunting Apes: Meat Eating and the Origins of Human Behavior* (Princeton, NJ: Princeton University Press, 1999); C. B. Stanford and H. T. Bunn, *Meat-eating and Human Evolution* (New York: Oxford University Press, 2001); A. D. Pfefferle et al., "Comparative Expression Analysis of the Phosphocreatine Circuit in Extant Primates: Implications

for Human Brain Evolution," *Journal of Human Evolution* 60, no. 2 (2011): 205–12.

9. See, for example, R. Trostle and R. Seeley, "Developing Countries Dominate World Demand for Agricultural Products," *Amber Waves,* August 5, 2013, http://www.ers.usda.gov/amber-waves/2013-august /developing-countries-dominate-world-demand-for-agricultural-pro ducts.aspx#.VPeFjPnF_QI; C. L. Delgado, "Rising Consumption of Meat and Milk in Developing Countries Has Created a New Food Revolution," *The Journal of Nutrition* 133, no. 11 (2003): 3907S-10S.

10. N. Fiala, "Meeting the Demand: An Estimation of Potential Future Greenhouse Gas Emissions from Meat Production," *Ecological Economics* 67, no. 3 (2008): 412–19; J. L. Capper, "Is the Grass Always Greener? Comparing the Environmental Impact of Conventional, Natural and Grass-Fed Beef Production Systems," *Animals* 2, no. 2 (2012): 127–43; J. L. Capper, "The Environmental Impact of Beef Production in the United States: 1977 Compared with 2007," *Journal of Animal Science* 89, no. 12 (2011): 4249–61; G. Boyd et al., "A 50-Year Comparison of the Carbon Footprint and Resource Use of the US Swine Herd: 1959–2009," report to the National Pork Board, May 22, 2012, http://research.pork.org/Results/ResearchDetail.aspx ?id=1574.

11. J. L. Capper and D. L. Hayes, "The Environmental and Economic Empact of Removing Growth-Enhancing Technologies from US Beef Production," *Journal of Animal Science* 90, no. 10 (2012): 3527. On safety see "Steroid Hormone Implants Used for Growth in Food-Producing Animals," U.S. Food and Drug Administration, October 14, 2014, http://www.fda.gov/AnimalVeterinary/SafetyHealth /ProductSafetyInformation/ucm055436.htm. See also K. P. Lone and L. A. van Ginkel, "Natural Sex Steroids and Their Xenobiotic Analogs in Animal Production: Growth, Carcass Quality, Pharmacokinetics, Metabolism, Mode of Action, Residues, Methods, and Epidemiology," *Critical Reviews in Food Science & Nutrition* 37, no. 2 (1997): 93–209; Joint Expert Committee on Food Additives (United Nations Food and Agriculture Organization and World Health Organization), "Residues of Some Veterinary Drugs in animals and Foods," FAO Food and Nutrition Paper, 1988; E. Doyle, "Human Safety of Hormone Implants Used to Promote Growth in Cattle: A Review of the Scientific Literature," *Food Research Institute Briefings,* July 2000, https://fri.wisc.edu/files/Briefs_File/hormone.pdf. For information on safety of rBST in dairy production, see J. C. Juskevidi and C. G. Guyer, "Bovine Growth Hormone: Human Food Safety Evaluation," *Science* 249, no. 4971 (1990): 875–84. Use of antibiotics (as compared

to hormones or beta-agonists) for growth promotion is a more complex issue. Because there are rules on withdrawal times and residues, there are no acute safety risks from eating meat from animals administered with growth-promoting antibiotics. However, there is some concern about the development of antibiotic-resistant bacteria. These concerns are somewhat mitigated by the fact that different classes of antibiotics are used in animal production than are used in human medicine, and the fact that the evidence suggests that antibiotic resistance in humans has come about mainly from human (rather than animal) overuse of antibiotics. For references on the matter, see a recent report by the president's Council of Advisors on Science and Technology, https://www.whitehouse.gov/sites/default/files/microsites/ostp/PCAST/pcast_carb_report_sept2014.pdf or A. G. Mathew, R. Cissell, and S. Liamthong, "Antiobiotic Resistance in Bacteria Associated with Food Animals: A United States Perspective of Livestock Production," *Foodborne Pathogens and Disease* 4, no. 2 (2007): 115–133. Or *The Use Of Drugs In Food Animals: Benefits and Risks* (Washington: National Academy Press, 1999) http://www.nap.edu/read/5137/chapter/1.

12. For a general discussion of animal welfare issues, see the book I wrote with Bailey Norwood, *Compassion by the Pound: The Economics of Farm Animal Welfare* (New York: Oxford University Press, 2011).

13. The cow-machine analogy isn't perfect because the cow is a living, feeling being and not a lifeless machine, but Post's lab-grown meat actually puts us even closer to the ideas in the thought experiment.

14. P. C. West et al., "Leverage Points for Improving Global Food Security and the Environment," *Science* 345, no. 6194 (July 18, 2014): 325–28.

15. For beef prices see Bureau of Labor Statistics, U.S. Department of Labor, Consumer Price Index Statistics, http://www.bls.gov/data; In 2014 the average price in a U.S. city for a pound of ground chuck (100 percent beef) was $4.

16. See Modern Meadow's FAQ at http://modernmeadow.com/ or Impossible Foods's FAQ at http://impossiblefoods.com/ for more information.

17. However, I note that several consumer surveys suggest shoppers remain a bit squeamish about the prospect of eating lab-grown meat. In a nationwide survey I fielded in 2014, I found that only about 19 percent of respondents said they would eat "a hamburger from meat grown in a lab." J. L. Lusk and S. Murray, "Food Demand Survey," Department of Agricultural Economics, Oklahoma State University, volume 2, issue 8, December 16, 2014, http://agecon.okstate.edu/faculty/publications/4950.pdf.

Chapter 8: Sustainable Farming

1. The university now has a sustainability website with links to efforts and activities all over campus: https://sustainability.okstate.edu/. Some of the results of the effort are recounted in "OSU Nabs Top 10 Finish in National Campus Conservation Competition," http://news .okstate.edu/articles/osu-nabs-top-10-finish-national-campus-con servation-competition?utm_source=OSU+Communications&utm _campaign=0059047f62-Headlines_5_21_155_20_2015&utm _medium=email&utm_term=0_ef85cdae6a-0059047f62-20596565.

2. "Lisa Turner on Organic Farming," *EconTalk,* hosted by Russ Roberts, December 24, 2012, http://www.econtalk.org/archives/2012/12/lisa _turner_on.html.

3. B. Tonn, A. Hemrick, and F. Conrad, "Cognitive Representations of the Future: Survey Results," *Futures* 38, no. 7 (2006): 810–29.

4. Discussion of the Magruder plots in this chapter relies on Bill Raun, interview by author, Stillwater, Oklahoma June 3, 2015; R. K. Boman et al., *The Magruder Plots: A Century of Wheat Research in Oklahoma* (Stillwater: Department of Agronomy, Oklahoma State University, 1996); K. Girma et al., "The Magruder Plots: Untangling the Puzzle," *Agronomy Journal* 99 (2007): 1191–98; R. W. Mullen et al., "The Magruder Plots—Long-Term Wheat Fertility Research," *Better Crops* 85, no. 4 (2001): 6–8; B. Arnall, "The Magruder Plots: 120 Years of Continuous Winter Wheat Research," *Better Crops with Plant Food* 97, no. 2 (2013): 29–31.

5. Quotes are from R. K. Boman, S. L. Taylor, W. R. Raun, G. V. Johnson, D. J. Bernardo, L. L. Singleton, *The Magruder Plots: A Century of Wheat Research in Oklahoma* (Stillwater: Oklahoma State University Press, 1996).

6. T. Hager, *The Alchemy of Air: A Jewish Genius, a Doomed Tycoon, and the Scientific Discover That Fed the World but Fueled the Rise of Hitler* (New York: Broadway Books, 2008), 11. The story about the accidental discovery of nitrogen fertilizer in South America is also from Hager.

7. R. K. Boman, S. L. Taylor, W. R. Raun, G. V. Johnson, D. J. Bernardo, L.L. Singleton, *The Magruder Plots: A Century of Wheat Research in Oklahoma* (Stillwater: Oklahoma State University Press, 1996), 7.

8. Data available from National Agricultural Statistics Service, U.S. Department of Agriculture: http://www.nass.usda.gov/Quick_Stats/.

9. R. K. Boman, S. L. Taylor, W. R. Raun, G. V. Johnson, D. J. Bernardo, L. L. Singleton, *The Magruder Plots: A Century of Wheat Research in Oklahoma* (Stillwater: Oklahoma State University Press, 1996), 24.

10. K. Girma et al., "The Magruder Plots: Untangling the Puzzle," *Agronomy Journal* 99 (2007): 1191–98. Quote is on page 1191.

11. K. L. Martin et al., "Plant-to-Plant Variability in Corn Production," *Agronomy Journal* 97, no. 6 (2005): 1603–11.

12. For more information about this device, see Ruan's website: http://www.nue.okstate.edu/Hand_Planter.htm.

13. Food and Agriculture Organization of the United Nations, "Maize, Rice, Wheat Farming Must Become More Sustainable," press release, December 19, 2014, http://www.fao.org/news/story/en/item/273303/icode/.

14. R. K. Boman, S. L. Taylor, W. R. Raun, G. V. Johnson, D. J. Bernardo, L. L. Singleton, *The Magruder Plots: A Century of Wheat Research in Oklahoma.* (Stillwater: Department of Agronomy, Oklahoma State University, 1996), 6.

15. Brett Carver, interview by author, June 10, 2015, Stillwater, OK.

16. National Agricultural Statistics Service, U.S. Department of Agriculture, "Duster Remains Top Wheat Variety for Fourth Year Running," *Oklahoma Wheat Variety Report,* March 2015 (crop year 2015), http://www.nass.usda.gov/Statistics_by_State/Oklahoma/Publications/Oklahoma_Crop_Reports/2015/ok_wheat_variety_2015.pdf.

17. Colorado Wheat, "Why Is the Wheat Genome So Complicated?" *Colorado Wheat Blog,* November 15, 2013, http://coloradowheat.org/2013/11/why-is-the-wheat-genome-so-complicated/.

18. Economic Research Service, U.S. Department of Agriculture, "U.S. Wheat Trade," June 22, 2015, http://www.ers.usda.gov/topics/crops/wheat/trade.aspx.

19. S. R. Pomeroy, "Gluten Intolerance May Not Exist," *Forbes,* May 15, 2014, http://www.forbes.com/sites/rosspomeroy/2014/05/15/non-celiac-gluten-sensitivity-may-not-exist/.

20. One such process is described in: K. Kempe, M. Rubtsova, and M. Gils, "Split-gene system for hybrid wheat seed production," *Proceedings of the National Academy of Sciences* 111(25) (2014): 9097–9102.

21. Data available from USDA NASS: http://www.nass.usda.gov/Quick_Stats/ or National Agricultural Statistics Service, U.S. Department of Agriculture, *Crop Production: Historical Track Records,* April 2014, http://www.nass.usda.gov/Publications/Todays_Reports/reports/croptr14.pdf.

Chapter 9: Waste Not, Want Not

1. J. C. Buzby, H. F. Wells, and J. Hyman, *The Estimated Amount, Value, and Calories of Postharvest Food Losses at the Retail and Consumer Levels in the United States,* EIB-121 (Washington, DC: Economic

Information Service, U.S. Department of Agriculture, February 2014), http://www.ers.usda.gov/publications/eib-economic-informa tion-bulletin/eib121.aspx.

2. BPI corporate history at http://www.beefproducts.com/history.php.

3. See material about Swift in the transcript of part 2 of *Chicago: City of the Century,* a 2003 film in the American Experience series of PBS, http://www.pbs.org/wgbh/amex/chicago/filmmore/pt_2.html. The film was produced and directed by Austin Hoyt for WGBH Boston.

4. M. Ogle, *In Meat We Trust: An Unexpected History of Carnivore America* (New York: Houghton Mifflin Harcourt, 2013).

5. The refrigerator worked this way: By heating a refrigerant with a low boiling point, the kerosene flame extracted heat from the refrigerator compartment, lowering the temperature to a point where spoilage could be slowed or stopped.

6. We conducted consumer sensory experiments involving blind taste tests of burgers that did and did not have finely textured beef. The product we actually tested was not BPI's but rather a competitor's product, but the basic processes used to make the two products are quite similar.

7. These figures are based on my extrapolations from the data in table 1 in J. R. Pruitt and D. P. Anderson, "Assessing the Impact of LFTB in the Beef Cattle Industry," *Choices,* fourth quarter 2012, http:// www.choicesmagazine.org/choices-magazine/theme-articles/pink -slimemarketing-uncertainty-and-risk-in-the-24-hour-news-cycle /assessing-the-impact-of-lftb-in-the-beef-cattle-industry.

8. Temple Grandin, interview by Indre Viskontas for *Point of Inquiry,* August 27, 2012, http://www.pointofinquiry.org/temple_grandin _the_science_of_livestock_animal_welfare/.

9. "LFTB Frequently Asked Questions," Beef Is Beef, a BPI website, http://beefisbeef.com/lftb-faq#lftb-source.

10. Annys Shin, "Engineering a Safer Burger," *The Washington Post,* June 12, 2008, http://www.washingtonpost.com/wp-dyn/content/article /2008/06/11/AR2008061103656.html. See BPI's website for a list of its and Roth's awards, http://www.beefproducts.com/awards_recog nition.php.

11. D. Rudman et al., "Ammonia Content of Food," *The American Journal of Clinical Nutrition* 26, no. 5 (1973): 487–90.

12. Sodium chloride and salt are, of course, the same thing. Yet, most consumers seem not to know. In a survey of more than a thousand Americans, we found that 66 percent of public believes added salt is natural, but only 32 percent believe added sodium chloride is natural. J. L. Lusk and S. Murray, *Food Demand Survey.* 1, no. 2 (June 14, 2013), http://agecon.okstate.edu/faculty/publications/4559.pdf.

13. The episode is from season 2, episode 1 of *Jamie Oliver's Food Revolution* entitled "Maybe L.A. Was a Big Mistake." First aired on April 12, 2011. The quotes represent the authors' transcription obtained from watching a recording of the episode.

14. N. Donley, "In Defense of Food Safety Leadership," *Food Safety News*, March 17, 2012, http://www.foodsafetynews.com/2012/03/in-defense-of-food-safety-leadership/#.VIXwSTHF98E; B. Gruley and E. Campbell, "The Sliming of Pink Slime's Creator," *Bloomberg Business Week*, April 12, 2012, http://www.businessweek.com/articles/2012-04-12/the-sliming-of-pink-slimes-creator.

15. *The Daily Show with Jon Stewart*, March 28, 2012, http://thedailyshow.cc.com/episodes/ag4oza/march-28—2012—ahmed-rashid.

Chapter 10: Food Bug Zappers

1. Raw milk is easier to find in France than it is in the United States, where its sale is restricted or banned in many states. Whether these restrictions are good or justifiable is a contentious issue that could require a book of its own. What is relevant here is that Pasteur invented a process that proved undeniably effective in reducing some forms of food-borne illness.

2. A. J. Langer et al., "Nonpasteurized Dairy Products, Disease Outbreaks, and State Laws—United States, 1993–2006," *Emerging Infectious Diseases* 18, no. 3 (2012): 385–91, http://www.cdc.gov/foodsafety/rawmilk/nonpasteurized-outbreaks.html.

3. A. L. Olmstead and P. W. Rhode, "An Impossible Undertaking: The Eradication of Bovine Tuberculosis in the United States," *Journal of Economic History* 64, no. 3 (September 2004), 734-772.

4. Reay Tannahill, *Food in History* (New York: Three Rivers, 1988), 64.

5. J. L. Lusk and S. Murray, *Food Demand Survey (FooDS)* 1, no. 2 (June 14, 2013), http://agecon.okstate.edu/faculty/publications/4559.pdf.

6. The quotes are, respectively, from the author's interviews with Stan Bailey and Frank Yiannas (see notes 18 and 23).

7. "Opinion of the Scientific Panel on Contaminants in the Food Chain on a request from the European Commission to Perform a Scientific Risk Assessment on Nitrate in Vegetables," *The EFSA Journal* 689 (2008): 1–79, http://www.efsa.europa.eu/sites/default/files/scientific_output/files/main_documents/contam_ej_689_nitrate_en.pdf.

8. "Estimating Foodborne Illness: An Overview," Centers for Disease Prevention and Control, April 17, 2014, http://www.cdc.gov/foodborneburden/; "Questions and Answers: 2011 Estimates," Centers for Disease Prevention and Control, June 27, 2013, http://www.cdc.gov/foodborneburden/questions-and-answers.html.

9.   S. Hoffmann, M. Batz, and J. G. Morris Jr., "Annual Cost of Illness-and Quality-Adjusted Life-Year Losses in the United States Due to 14 Foodborne Pathogens," *Journal of Food Protection* 75, no. 7 (2012): 1291–1302; R. Scharff, "Economic Burden from Health Losses Due to Foodborne Illness in the United States," *Journal of Food Protection* 75, no. 1 (2012): 123–31; S. Hoffman and T. D. Anekwe, "Making Sense of Recent Cost-of-Foodborne-Illness Estimates," *Economic Information Bulletin,* no. 118, September 2013, http://www.ers.usda.gov/publications/eib-economic-information-bulletin/eib118.aspx.

10.   M. R. Thomsen and A. M. McKenzie, "Market Incentives for Safe Foods: An Examination of Shareholder Losses from Meat and Poultry Recalls," *American Journal of Agricultural Economics* 83, no. 3 (2001): 526–38.

11.   M. R. Thomsen, R. Shiptsova, and S. J. Hamm, "Sales Responses to Recalls for Listeria Monocytogenes: Evidence from Branded Ready-to-Eat Meats," *Review of Agricultural Economics* 28, no. 4 (2006): 482–93.

12.   M. Basu, "Unprecedented Verdict: Peanut Exec Found Guilty in Deadly Salmonella Outbreak," *CNN.com,* September 19, 2014, http://www.cnn.com/2014/09/19/us/peanut-butter-salmonella-trial/index.html.

13.   K. McCoy, "Peanut Exec in Salmonella Case Gets 28 Years." USA Today, September 22, 2015, http://www.usatoday.com/story/money/business/2015/09/21/peanut-executive-salmonella-sentencing/72549166/; D. Flynn, "Parnell Brothers Taken into Custody after Convictions," *Food Safety News,* September 20, 2014, http://www.foodsafetynews.com/2014/09/stewart-and-michael-parnell-booked-into-albany-ga-jail-after-convictions-in-pca-trial/#.VdsqYqSFOpo.

14.   D. Flynn, "Odwalla Apple Juice E. Coli Outbreak," *Food Safety News,* September 17, 2009, http://www.foodsafetynews.com/2009/09/meaningful-outbreak-4-odwalla-apple-juice-e-coli-o157h7-outbreak/#.Vd4SFPlVhBc.

15.   J. Klineman, "Expo East: High Pressure to Hit Naked, Odwalla Hard, Despite Family Ties," *BevNet,* September 30, 2013, http://www.bevnet.com/news/2013/expo-east-high-pressure-to-hit-naked-odwalla-hard-despite-family-ties/.

16.   "Kinetics of Microbial Inactivation for Alternative Food Processing Technologies—High Pressure Processing," U.S. Food and Drug Administration, December 18, 2014, http://www.fda.gov/Food/FoodScienceResearch/SafePracticesforFoodProcesses/ucm101456.htm.

17.   M. H. Kothary and U. S. Babu, "Infective Dose of Foodborne Pathogens in Volunteers: A Review," *Journal of Food Safety* 21, no. 1 (2001): 49–68.

18. Stan Bailey, telephone interview by author, August 19, 2015.

19. Yang Jie, "Is Your Food Safe? New 'Smart Chopsticks' Can Tell," *China Real Time* (blog), *Wall Street Journal China*, September 3, 2014, http://blogs.wsj.com/chinarealtime/2014/09/03/is-your-food-safe-baidus-new-smart-chopsticks-can-tell/; J. M. Azzarelli et al., "Wireless Gas Detection with a Smartphone via RF Communication," *Proceedings of the National Academy of Sciences* 111, no. 51 (2014): 18162–66.

20. C. Seigner, "Scientists Develop Handheld Device to Detect Bacteria on Food," *Food Safety News*, November 1, 2013, http://www.foodsafetynews.com/2013/11/personal-handheld-sensors-to-detect-pathogens-could-be-the-future/#.VdyTuvlVhBc; MS Tech Ltd., "Foodscan: Handheld Food Contamination Detector Using HF-QCM Sensor Technology," http://www.ms-technologies.com/products/foodscan/; K. Monks, "This 'Star Trek' Style Molecular Sensor Fits in Your Hand, Reads Your Food," *CNN.com*, May 2, 2014, http://www.cnn.com/2014/05/02/tech/innovation/molecular-sensor-fits-in-your-hand/; A. Pasolini, "Penguin Device Checks Your Food for Antiobiotic Residue," *gizmag*, June 27, 2014, http://www.gizmag.com/penguin-antibiotic-sensor/32719/.

21. "Estimating Foodborne Illness."

22. Frank Yiannas, telephone interview by author, August 10, 2015

23. M. F. Bellemare, R. P. King, and N. Nguyen "Farmers Markets and Food-Borne Illness," working paper, July 17, 2015, http://marcfbellemare.com/wordpress/wp-content/uploads/2015/07/BellemareKingNguyenFarmersMarketsJuly2015.pdf.

24. C. E. Park and G. W. Sanders, "Occurrence of Thermotolerant Campylobacters in Fresh Vegetables Sold at Farmers' Outdoor Markets and Supermarkets," *Canadian Journal of Microbiology* 38, no. 4 (1992): 313–16; J. Scheinberg, S. Doores, and C. N. Cutter, "A Microbiological Comparison of Poultry Products Obtained from Farmers' Markets and Supermarkets in Pennsylvania," *Journal of Food Protection* 33, no. 3 (2013): 259–64.

25. J. Crews, "Walmart Poultry Safety Campaign Takes a Page from 2010 Beef Initiative," *Meat+Poultry*, January 12, 2015, http://www.meatpoultry.com/articles/news_home/Food_Safety/2015/01/Walmart_poultry_safety_campaig.aspx?ID={AF6B7C1B-64B2-4637-93A1-9DC5DAC15963}&cck=1.

26. B. Salvage, "Walmart's Yiannis Details Food-Safety Initiative," *Meat+Poultry*, December 18, 2014, http://www.meatpoultry.com/articles/news_home/Business/2014/12/Walmarts_Yiannis_details_food.aspx?ID=%7B6ABB75CA-4650-4B3D-AC86-7387BF9147C6%7D&cck=1.

27. P. Crandall et al., "Companies' Opinions and Acceptance of Global Food Safety Initiative Benchmarks After Implementation," *Journal of Food Protection* 75, no. 9 (2012): 1660–72.

28. L. K. Strawn et al., "Big Data in Food," *Food Technology* 69, no. 2 (2015): 42–49.

Chapter 11: The Case for Food and Agricultural Innovation

1. Jules B. Billard, "The Revolution in American Agriculture," *National Geographic,* February 1970, 147–185.

2. J. Foley, "A Five-Step Plan to Feed the World," *The Future of Food,* special issue, *National Geographic,* 2014, http://www.nationalgeo graphic.com/food-special-compilation/.

3. R. Hoppe, *Structure and Finances of U.S. Farms: Family Farm Report,*" EIB-132, (Washington, DC: Economic Information Service, U.S. Department of Agriculture, December 2014), http://www.ers.usda .gov/publications/eib-economic-information-bulletin/eib132.aspx.

4. G. Dickie, "Q&A: Inside the World's Largest Farm," *National Geographic,* July 19. 2014, http://news.nationalgeographic.com/news/2014 /07/140717-japan-largest-indoor-plant-factory-food/; S. N. Bhanoo, "Vertical Farms Will be Big, but for Whom?" *Fast Company,* December 3, 2014, http://www.fastcompany.com/3039087/elasticity /vertical-farms-will-be-big-but-for-who.

5. Economic Research Service (ERS), U.S. Department of Agriculture, "Agricultural Productivity in the U.S.," table 1 in "National Tables 1948–2011, June 4, 2015, http://www.ers.usda.gov/data-products /agricultural-productivity-in-the-us.aspx. Herbicide quantities are based on data related to pounds of active ingredient J. Fernandez-Cornejo et al., *Pesticide Use in U.S. Agriculture: 21 Selected Crops, 1960–2008,*" EIB-124 (Washington, DC: Economic Information Service, U.S. Department of Agriculture, May 2014), table 2, http:// www.ers.usda.gov/media/1424185/eib124.pdf. Toxicity is from figure 4.1, p. 28.

6. ERS, "Agricultural Productivity in the U.S." This particular data set shows an increase in use of pesticides over time, but the analysis in Fernandez-Cornejo et al., *Pesticide Use in U.S. Agriculture,* suggests this is the result of using "quality-adjusted" quantities. Because pesticide quality has improved over time, farmers now use less than they would have if quality had not improved.

7. D. Jorgenson, F. Gollop, and B. Fraumeni, *Productivity and U.S. Economic Growth* (Cambridge, MA: Harvard University Press, 1987); D. Jorgenson, M. Ho, and K. Stiroh, *Productivity: Information Technology and the American Growth Resurgence* (Cambridge,

MA: MIT Press, 2005); Natural Resources Conservation Service, U.S. Department of Agriculture, "Soil Erosion on Cropland 2007," http://www.nrcs.usda.gov/wps/portal/nrcs/detail/national/technical /nra/nri/?cid=stelprdb1041887; J. Horowitz, R. Ebel, and K. Ueda, *"No-Till" Farming Is a Growing Practice,* EIB-70 (Washington, DC: Economic Information Service, U.S. Department of Agriculture, November 2010), http://www.ers.usda.gov/media/135329/eib70.pdf; J. Fernandez-Cornejo et al., *Genetically Engineered Crops in the United States,* ERR-162, (Washington, DC: Economic Research Service, U.S. Department of Agriculture, February 2014); R. Myers, *Cover Crop Trends in the U.S.,* February 2015, http://css.wsu.edu/biofuels /files/2015/02/MyersCoverCrops2015OSDS.pdf.

8. For a broader discussion of these issues, more citations, and discussion of data, see J. L. Lusk, "Role of Technology in the Global Economic Importance and Viability of Animal Protein Production," *Animal Frontiers* 3 (2013): 20–27.

9. J. L. Capper, "The Environmental Impact of Beef Production in the United States: 1977 Compared with 2007," *Journal of Animal Science* 89 (2011): 4249–61. Capper estimates the water, carbon, and waste impacts associated with producing a billion kilograms of beef in 1977 and 2007. In 2014 the United States produced about eleven billion kilograms of beef.

10. J. Foley, "A Five-Step Plan." 13.

11. For more about Steven Pinker, see http://cultivatingthought.com /author/steven-pinker/.

12. N. Johnson, "Would You Like Some Criticism on Your GMO-Free Chipotle Burrito?" *grist,* May 7, 2015, http://grist.org/food /would-you-like-some-criticism-on-your-gmo-free-chipotle-burrito/.

13. S. J. Dubner, "Waiter, There's a Physicist in My Soup," pt. 1, *Freakonomics,* January 27, 2011, http://freakonomics.com/2011/01/27 /freakonomics-radio-waiter-theres-a-physicist-in-my-soup-part-i/; Alice Waters, *The Art of Simple Food: Notes, Lessons, and Recipes from a Delicious Revolution* (New York: Clarkson Potter, 2007), e-book edition.

14. S. C. P. Williams, "Humans May Harbor More Than 100 Genes from Other Organisms," *Science,* March 12, 2015, http://news.sciencemag .org/biology/2015/03/humans-may-harbor-more-100-genes-other -organisms; Peter Singer, interview by Krishna Chokshi, *Brown Journal of World Affairs* 14, no. 1 (Fall–Winter 2007):. 135–43.136

15. R. Laudan, "A Plea for Culinary Modernism," *Jacobin,* May 22, 2015, https://www.jacobinmag.com/2015/05/slow-food-artisanal-natural -preservatives/.

16. Ibid.

17. J. L. Lusk and S. Murray, *Food Demand Survey (FooDS)* 2, no. 8 (December 16, 2014), agecon.okstate.edu/faculty/publications/4950 .pdf. The issues, in order of importance, were: having affordable food for me and my family (23.2 percent); changing the type and quantity of food eaten to address obesity, diabetes, and heart disease (12 percent); producing enough food to meet the demands of a growing world population (11.1 percent); finding ways to prevent adverse environmental impacts of food production (10 percent); current government policies directed at farms and food production (9.5 percent); profitability of U.S. farmers (9 percent); involvement of large corporations in agriculture, food production, and food processing (8.5 percent); U.S. food imports and exports (8.4 percent); and inequitable distribution of food through the world (8.1 percent).

18. J. Wesseler, S. Kaplan, and D. Zilberman, "The Cost of Delaying Approval of Golden Rice," *Agricultural Resource Economics Update* 17, no. 3 (2014): 1–3, http://library.wur.nl/WebQuery/wurpubs/453391.

19. Peter Singer, interview by L. Mensell and M. Tholl, *The World Post,* November 26, 2014, http://www.huffingtonpost.com/2014/10/30 /peter-singer-cautious-caution_n_6075116.html; Peter Singer, "Scientists Playing God Will Save Lives," *The Guardian,* June 13, 2010, http://www.theguardian.com/commentisfree/2010/jun/13/science -playing-god-climate-change.

20. Data on cattle inventory by state are available from USDA National Agricultural Statistical Service: http://www.nass.usda.gov /Quick_Stats/.

21. A. L. Olmstead and P. W. Rhode, "Reshaping the Landscape: The Impact and Diffusion of the Tractor in American Agriculture, 1910–1960." *Journal of Economic History* 61 (2001): 663–98.

22. For current data on agricultural employment, see "Employment Projections: Employment by Major Industry Sector," Bureau of Labor Statistics, U.S. Department of Labor, December 2013, http:// www.bls.gov/emp/ep_table_201.htm. For historical agricultural employment data and white-collar jobs, see Q. Bui, "How Machines Destroy (and Create!) Jobs in 4 Graphs," *NPR.org,* May 18, 2015, http://www.npr.org/sections/money/2015/05/18/404991483 /how-machines-destroy-and-create-jobs-in-4-graphs.

23. Michael Pollan, interview by Joe Fassler, *The Atlantic,* April 23, 2013, http://www.theatlantic.com/entertainment/archive/2013/04/the -wendell-berry-sentence-that-inspired-michael-pollans-food-obses sion/275209/.

24. D. Sax, *The Tastemakers: Why We're Crazy for Cupcakes but Fed Up with Fondue* (New York: Public Affairs, 2014), 112.

25. For trends in leisure see M. Aguiar and E. Hurst, "Measuring Trends in Leisure: The Allocation of Time over Five Decades," *Quarterly Journal of Economics* 122 (2007: 969–1006. For data on unemployment after 1940 see "Labor Force Statistics from the Current Population Survey," Bureau of Labor Statistics, U.S. Department of Labor, February 12, 2015, http://www.bls.gov/cps/cpsaat01.htm. Older unemployment data appear in C. Romer, "Spurious Volatility in Historical Unemployment Data," *Journal of Political Economy* 94 (1986): 1–37.

26. Bureau of Labor Statistics, "Employment Projections."

27. Hoppe, "Structure and Finances of U.S. Farms."

28. B. M. Clark et al., "The Role of an Agribusiness Index in a Modern Portfolio," *Agricultural Finance Review* 72, no. 3 (2012): 362–80; H. O. Zapata, J. D. Detre, and T. Hanabuchi, "Historical Performance of Commodity and Stock Markets," *Journal of Agricultural and Applied Economics* 44, no. 3 (2012): 339–57; J. M. D'Antoni and D. J. Detre, "Are Agribusiness Stocks an Investor Safe Haven?" *Agricultural Finance Review* 74, no. 4 (2014): 522–38.

29. See tables 7 and 8 and the table entitled "Percent of consumer expenditures spent on food, alcoholic beverages, and tobacco that were consumed at home, by selected countries, 2014," http://www.ers.usda.gov/data-products/food-expenditures.aspx.

30. U.S. Department of Agriculture, *FY 2016 USDA Budget Summary and Annual Performance Plan,* 121, http://www.obpa.usda.gov/budsum/fy16budsum.pdf.

31. "What We Do," Agricultural Research Service, U.S. Department of Agriculture, http://www.ars.usda.gov/AboutUs/AboutUs.htm.

32. USDA, *FY 2016 USDA Budget Summary,* 121. http://www.obpa.usda.gov/budsum/fy16budsum.pdf.

33. These are the amounts allocated to the OSU Oklahoma Agricultural Experiment Station and OSU Agricultural Extension Service through the Regents for Higher Education in fiscal year 2013–14. See *State of Oklahoma Executive Budget for the Fiscal Year Ending June 30, 2015,* C-24, http://www.ok.gov/OSF/documents/bud15.pdf. The experiment station and extension service receive additional state monies through other agencies and programs; the extension service is also funded by county governments.

34. K. O. Fuglie and P. W. Heisey, "Economic Returns to Public Agricultural Research," Economic Brief no. 10, September 2007; J. M. Alston et al., "The Economic Returns to U.S. Public Agricultural Research," *American Journal of Agricultural Economics* 93, no. 5 (2011): 1257–77.

35. J. M. Alston, J. M. Beddow, and P. G. Pardey, 2009. "Agricultural Research, Productivity, and Food Prices in the Long Run," *Science,*

September 4, 2009, 1209–10; K. P. Fuglie et al., "The Contribution of Private Industry to Agricultural Innovation," *Science,* November 23, 2012, 1031–32.

36. Bill and Melinda Gates Foundation, *Who We Are: Annual Report 2013,* http://www.gatesfoundation.org/Who-We-Are/Resources-and -Media/Annual-Reports/Annual-Report-2013.

37. In the interest of full disclosure, I hold a fellowship through the Oklahoma Council of Public Affairs that is partially funded by the Noble Foundation.

38. Quotes from Buckner come from an email exchange with the author on May 26, 2015.

39. Budget information is from: Samuel Roberts Noble Foundation, *The Enduring Lessons of Lloyd Noble: 2012 Annual Report,* http://www .noble.org/Global/news/annualreport/2012/annualreport.pdf.

Index